REVISE EDEXCEL AS/A LEVEL
Chemistry

D1586845

REVISION GUIDE

Series Consultant: Harry Smith

Author: Nigel Saunders

A note from the publisher

In order to ensure that this resource offers high-quality support for the associated Pearson qualification, it has been through a review process by the awarding body. This process confirms that this resource fully covers the teaching and learning content of the specification or part of a specification at which it is aimed. It also confirms that it demonstrates an appropriate balance between the development of subject skills, knowledge and understanding, in addition to preparation for assessment.

Endorsement does not cover any guidance on assessment activities or processes (e.g. practice questions or advice on how to answer assessment questions), included in the resource nor does it prescribe any particular approach to the teaching or delivery of a related course.

While the publishers have made every attempt to ensure that advice on the qualification and its assessment is accurate, the official specification and associated assessment guidance materials are the only authoritative source of information and should always be referred to for definitive guidance.

Pearson examiners have not contributed to any sections in this resource relevant to examination papers for which they have responsibility.

Examiners will not use endorsed resources as a source of material for any assessment set by Pearson.

Endorsement of a resource does not mean that the resource is required to achieve this Pearson qualification, nor does it mean that it is the only suitable material available to support the qualification, and any resource lists produced by the awarding body shall include this and other appropriate resources.

Contents

1-to-1
page match with the
Chemistry Revision
Workbook
ISBN 9781447989943

A small bit of small print
Edexcel publishes Sample Assessment Material and the Specification on its website. This is the official content and this book should be used in conjunction with it. The questions in *Now try this* have been written to help you practise every topic in the book. Remember: the real exam questions may not look like this.

Atomic structure and isotopes

You need to know the structure of an atom and the properties of the subatomic particles it contains. You must also be able to calculate the number of these particles from given data.

Structure of an atom

An atom comprises:

- a nucleus containing protons and (almost always) neutrons

- one or more electrons in energy levels around the nucleus.

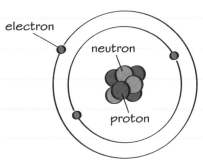

Properties of subatomic particles

Make sure you know these properties:

Particle	proton	neutron	electron
Symbol	p	n	e^-
Relative mass	1	1	$\frac{1}{1840}$
Relative charge	+1	0	−1

The masses and charges of the neutron and electron are compared to those of the proton. The + and − signs in the relative charges are needed. Remember:

- positive proton   • neutral neutron.

Full chemical symbols

These provide information about the number of particles in the nucleus.

$$\text{mass number} \rightarrow {}^{A}_{Z}X \leftarrow \text{symbol}$$
$$\text{atomic number} \rightarrow$$

The carbon-12 atom, ${}^{12}_{6}C$, is the standard atom. Its relative isotopic mass is defined as exactly 12. ${}^{14}_{6}C$ is another isotope of carbon – its relative isotopic mass is 14.

Some key definitions

✓ **Mass number**, A is the number of protons added to the number of neutrons in the nucleus.

✓ **Atomic number**, Z is the number of protons in the nucleus.

✓ **Relative isotopic mass** is the mass of an atom of an isotope compared to $\frac{1}{12}$th the mass of a ${}^{12}C$ atom.

Be careful! The top numbers in the Periodic Table given in the examination are relative atomic masses, A_r (not mass numbers).

Worked example

(a) State, with reference to particles, the meaning of the term isotopes. **(2 marks)**

Atoms that have the same number of protons but different numbers of neutrons.

(b) Calculate the number of each type of particle present in a ${}^{23}_{11}Na^+$ ion. **(3 marks)**.

11 protons, 11 − 1 = 10 electrons
(23 − 11) = 12 neutrons

Isotopes can also be defined as atoms with the same atomic number but different mass numbers. This definition would not be an appropriate answer here as the question asks you to refer to particles.

The bottom number gives the number of protons. This is also the number of electrons in a neutral atom. However, for an ion, remember to:

- subtract electrons (for positive ions)
- add electrons (for negative ions).

$$\text{Number of neutrons} = \frac{\text{mass number}}{\text{(top number)}} - \frac{\text{atomic number}}{\text{(bottom number)}}$$

Now try this

1 An atom contains two more protons than, but the same number of neutrons as, an atom of ${}^{18}O$. Deduce the symbol, including the mass number and the atomic number, of this atom. **(2 marks)**

2 Calculate the number of each type of particle present in a ${}^{31}_{15}P^{3-}$ ion. **(3 marks)**

Use the Periodic Table to identify the element once you know its atomic number.

3 (a) Explain why ${}^{1}H$ and ${}^{2}H$ are isotopes of the same element. **(2 marks)**

(b) Identify the particle present in an atom of ${}^{2}H$ but absent in an atom of ${}^{1}H$. **(1 mark)**

4 (a) Define the term atomic number, and explain why an element's atoms may have different mass numbers. **(2 marks)**

(b) Explain why an atom is electrically neutral overall. **(2 marks)**

Mass spectrometry

A mass spectrometer allows you to measure the characteristics of molecules by converting them to ions, which can be manipulated by electric and magnetic fields.

Relative abundance against m/z ratio

The mass spectrometer produces positive ions from gaseous samples injected into it.

A **mass spectrum** is a chart that shows:

- the relative abundance of each ion
- the mass/charge ratio (m/z) of each ion.

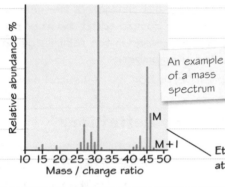

An example of a mass spectrum

Ethanol produces a molecular ion peak M at m/z = 46.0, so its M_r is 46.0 too.

Finding relative molecular mass, M_r

The peak with the highest mass/charge ratio represents the **molecular ion**:

- It is the peak furthest to the right.
- Its m/z ratio is equal to the M_r.

Isotopes such as ^{13}C may cause a very small peak at M+1, which can be ignored when finding M_r.

Be careful! You should use the term relative formula mass instead of relative molecular mass for compounds with giant structures (metals, ionic compounds and giant molecules (e.g. SiO_2)).

Worked example

The mass spectrum of boron is shown below.

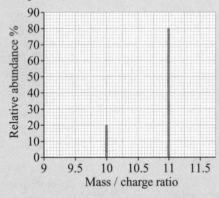

(a) Explain why there are two peaks.   **(1 mark)**

Boron has two isotopes.

Maths skills Pay careful attention to the scale on graphs. On the vertical axis 5 small squares represent 10% so 1 small square represents 2%

Relative atomic mass A_r is the weighted mean mass of an atom of an element compared to $\frac{1}{12}$th the mass of a ^{12}C atom. A_r is a quantity with no units.

(b) Calculate the relative atomic mass, A_r, of boron.   **(2 marks)**

$$\frac{(10 \times 20) + (11 \times 80)}{100} = \frac{200 + 880}{100} = 10.8$$

A sample of bromine contains two isotopes, ^{79}Br and ^{81}Br. Its mass spectrum has three peaks close together.

(c) Identify the species that give rise to peaks at m/z 158 and 162.   **(2 marks)**

$(^{79}Br-^{79}Br)^+$ and $(^{81}Br-^{81}Br)^+$

(d) State the m/z value of the other peak, and identify the species that gives rise to it.   **(2 marks)**

m/z = 160, caused by $(^{79}Br-^{81}Br)^+$

Remember that the peaks in mass spectra are caused by positively charged ions, so a single positive charge must be shown in your answers.

Now try this

1 Define the term relative atomic mass.   **(2 marks)**

2 The element chlorine has two stable isotopes, ^{35}Cl and ^{37}Cl. Describe the peaks seen in the mass spectrum of chlorine, due to Cl_2^+ ions. Explain your answer.   **(3 marks)**

Shells, sub-shells and orbitals

Electrons surround the nucleus of an atom and are arranged in orbitals, sub-shells and shells.

Quantum shells

Electrons in atoms exist in energy levels called **quantum shells**.

shell 4 ────
shell 3 ────
             increasing energy
             increasing distance
shell 2 ────

shell 1 ──── closest to nucleus

Sub-shells and orbitals

Each shell contains one or more **sub-shells**, which have the letters s, p or d.

Each sub-shell contains different numbers of **orbitals**.

Sub-shell	s	p	d	f
Number of orbitals	1	3	5	7
Number of electrons	2	6	10	14

☑ An orbital is a region around the nucleus where there is a high probability of finding an electron.

☑ An orbital can hold up to two electrons with opposite spins.

Worked example

Draw diagrams to show the shape of an s-orbital and of a p-orbital. **(2 marks)**

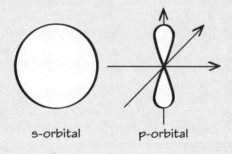

s-orbital   p-orbital

The arrows represent the three axes in space x, y and z. The p-orbital could have been drawn in one of the other two orientations instead.

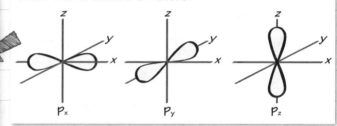

p_x p_y p_z

You must show an s-orbital as a circle. You do not need to know the shape of d- or f-orbitals.

Electrons in shells

You need to know the maximum number of electrons in the first four quantum shells.

Shell	Sub-shell(s)	Maximum number of electrons
4	4s 4p 4d 4f	2 + 6 + 10 + 14 = 32
3	3s 3p 3d	2 + 6 + 10 = 18
2	2s 2p	2 + 6 = 8
1	1s	2

The 4d and 4f sub-shells are included here only so you can see why the fourth shell can contain up to 32 electrons.

Electrons in orbitals

Electrons have a property called **spin**.

The electrons in an orbital have opposite spins.

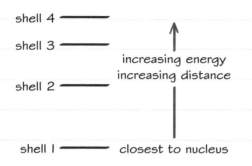

4p [↑↓][↑↓][↑↓]
3d [↑↓][↑↓][↑↓][↑↓][↑↓] ← 3d is higher in energy than 4s
4s [↑↓]
3p [↑↓][↑↓][↑↓]
3s [↑↓]
2p [↑↓][↑↓][↑↓] ← $2p_x$ $2p_y$ $2p_z$ orbitals
2s [↑↓]
1s [↑↓] ← arrows represent electrons with opposite spin

Now try this

1 (a) Explain what is meant by the term **orbital**. **(2 marks)**

   (b) Draw the shapes of an s-orbital and a p-orbital. **(2 marks)**

2 State the maximum number of electrons that can occupy:
   (a) an s-, a p- and a d-sub-shell **(1 mark)**
   (b) each of the first four quantum shells. **(1 mark)**

3 Explain why electrons may be represented as arrows in boxes. **(2 marks)**

Electronic configurations

The electronic configuration of an atom is its distribution of electrons among orbitals. You need to be able to predict the configuration of an atom from its atomic number (for $Z = 1$ to 36, i.e. H to Kr).

The '1s' notation

The electronic configuration of Ca, $Z = 20$, is:

the shell number →

$1s^2\ 2s^2\ 2p^6\ 3s^2\ 3p^6\ 4s^2$

← the number of electrons in the sub-shell

the orbital type

After calcium, 3d is lower in energy than 4s.

The electronic configuration of Sc, $Z = 21$, is:

$1s^2\ 2s^2\ 2p^6\ 3s^2\ 3p^6\ 3d^1\ 4s^2$

You could write $4s^2\ 3d^1$ here instead if you prefer.

The 'electrons in boxes' notation

Hund's rule: electrons occupy orbitals singly before pairing happens.

Pauli exclusion principle: electrons in the same orbital must have opposite spins.

For example, the electronic configuration for nitrogen is $1s^2\ 2s^2\ 2p_x^1\ 2p_y^1\ 2p_z^1$ (and not $1s^2\ 2s^2\ 2p_x^2\ 2p_y^1\ 2p_z^0$). It is shown as:

1s   2s $2p_x$ $2p_y$ $2p_z$

[↑↓] [↑↓] [↑] [↑] [↑]

Worked example

Complete the electronic configurations of:

(a) vanadium ($Z = 23$), $1s^2$ ...   **(1 mark)**

$1s^2\ 2s^2\ 2p^6\ 3s^2\ 3p^6\ 3d^3\ 4s^2$

(b) chromium ($Z = 24$), $1s^2$ ...   **(1 mark)**

$1s^2\ 2s^2\ 2p^6\ 3s^2\ 3p^6\ 3d^5\ 4s^1$

In chromium and copper atoms, one 4s electron is promoted to a 3d orbital.

Chromium is not: $1s^2\ 2s^2\ 2p^6\ 3s^2\ 3p^6\ 3d^4\ 4s^2$

Copper is:

$1s^2\ 2s^2\ 2p^6\ 3s^2\ 3p^6\ 3d^{10}\ 4s^1$ and not

$1s^2\ 2s^2\ 2p^6\ 3s^2\ 3p^6\ 3d^9\ 4s^2$

Blocks and groups in the Periodic Table

A **group** is a vertical column in the Periodic Table. The electronic configuration of the highest occupied shell is the same for all the atoms in a main group. This is why elements in a group have similar chemical properties.

n represents the shell number.

Group 0 is also called Group 8.

Group	1	2	3	4	5	6	7	0
Outer shell configuration	ns^1	ns^2	$ns^2\ np^1$	$ns^2\ np^2$	$ns^2\ np^3$	$ns^2\ np^4$	$ns^2\ np^5$	$ns^2\ np^6$

The Periodic Table can be divided into **blocks**:

- s-block elements have their last added electron in an s-orbital
- p-block elements have their last added electron in a p-orbital
- d-block elements have their last added electron in a d-orbital

The number of electrons in the outer shell equals the group number.

Hydrogen and helium are in the s-block, as their electron configurations are $1s^1$ and $1s^2$.

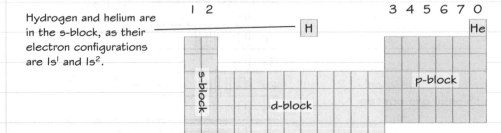

Now try this

1 Give the electronic structures of the following elements in s, p and d notation:

(a) silicon, $Z = 14$ **(1 mark)**

(b) iron, $Z = 26$ **(1 mark)**

(c) bromine, $Z = 35$ **(1 mark)**

2 (a) State the period, group and block for an atom with the electronic configuration $1s^2\ 2s^2\ 2p^4$.   **(1 mark)**

(b) Show the electronic configuration for this atom using the electrons in boxes notation.   **(1 mark)**

Ionisation energies

Trends in ionisation energies provide evidence for quantum shells and sub-shells.

Ionisation energy definitions

First ionisation energy is the energy per mole needed to remove an electron from gaseous atoms:

$$X(g) \rightarrow X^+(g) + e^-$$

More electrons can be removed from ions.

For example, **second ionisation energy** is the energy per mole needed to remove an electron from gaseous ions with a single positive charge:

$$X^+(g) \rightarrow X^{2+}(g) + e^-$$

The first and second ionisation energies are two of a series of **successive ionisation energies** until all the electrons are removed.

> **Be careful!** The second ionisation energy does *not* correspond to:
> $$X(g) \rightarrow X^{2+}(g) + 2e^-$$

Successive ionisation energies

Successive ionisation energies provide evidence for quantum shells: there is a large increase in ionisation energy from one shell to the next.

The graph is for sodium, $1s^2 2s^2 2p^6 3s^1$.

The first large increase is after one electron is removed, showing that sodium is in Group 1.

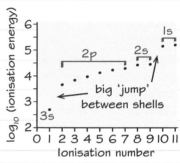

Maths skills Ionisation energies are shown here in logarithms to base 10. This means the power of 10 is plotted. This is because they are very large and cover several orders of magnitude.

First ionisation energy

This decreases going down a group:
- The energy of the sub-shell from which the electron is removed increases.
- The amount of shielding increases.

The number of protons also increases, but its effect is less significant than the other factors.

Shielding

Shielding is electron–electron repulsion. It exists between two electrons in the same orbital and between electrons in different orbitals.

Worked example

The graph shows how the first ionisation energy changes across Period 3. Use crosses to show the approximate values for the remaining elements and join the crosses. **(3 marks)**

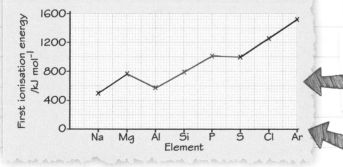

Evidence for sub-shells

The first ionisation energy for Al is lower than it is for Mg. This is because the 3p electron removed from Al is higher in energy than the 3s electron removed from Mg, and has greater shielding.

The key points on the graph are:
- Al is higher than Na but lower than Mg.
- Si is higher than Mg but lower than P.
- P is higher than S but lower than Cl.

You do not need to recall first ionisation energy values but you must be able to sketch the graph.

Now try this

The table shows all the successive ionisation energies for an element in the p-block.

Ionisation number	1st	2nd	3rd	4th	5th	6th	7th
Ionisation energy/kJ mol^{-1}	1402	2856	4578	7475	9445	53 268	64 362

(a) Explain which group this element belongs to. **(1 mark)**

(b) Which ionisation energies correspond to removing an electron from an s-orbital? **(2 marks)**

Periodicity

Periodicity is a repeating pattern of physical and chemical properties across different periods.

Atomic radius

The **atomic radius** of an element is the distance from the centre of the nucleus to the boundary of the electron cloud. It shows periodicity. Atomic radius:
- decreases across a period
- increases from one period to the next.

There is a negligible increase in shielding going across a period, but the nuclear charge increases. This increases the force of attraction between the nucleus and outer electrons.

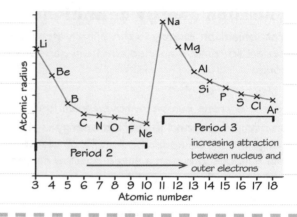

Melting and boiling temperatures

These vary across a period depending on the type of structure and bonding present.
- Metals have metallic bonding and a giant lattice structure.
- Boron, carbon and silicon have covalent bonding and a giant lattice structure.
- Noble gases in Group 0 are monatomic – they exist as individual atoms.
- Other non-metals have covalent bonding and a simple molecular structure.

Page 18 has more information about the relationship between structure and melting and boiling temperatures.

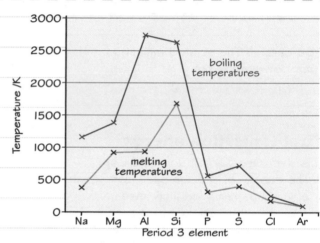

Worked example

(a) Complete the electronic configurations for phosphorus and sulfur. **(2 marks)**

		3s	3p		
P	[Ne]	↑↓	↑	↑	↑
S	[Ne]	↑↓	↑↓	↑	↑

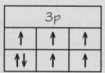

[Ne] represents the electronic configuration of neon, the noble gas before phosphorus and sulfur in the Periodic Table.

(b) Explain why the first ionisation energy of sulfur is lower than that of phosphorus. **(2 marks)**

In sulfur, an electron is removed from a 3p-orbital containing two electrons. These electrons repel each other more strongly than the ones in phosphorus, so the electron is lost more easily.

The answer could also have been given in terms of a 3p sub-shell, rather than an orbital.

Now try this

1 Describe the structure and bonding present in lithium, boron and oxygen. **(3 marks)**

2 What is the trend in melting temperature in Period 2 (A, B, C or D)? **(1 mark)**
   ☐ **A** A steady increase.
   ☐ **B** A steady decrease.
   ☐ **C** An increase to carbon then a decrease.
   ☐ **D** A decrease to carbon then an increase.

3 Which of the elements, gallium to strontium, is likely to have the smallest atomic radius? **(1 mark)**

4 Explain, in terms of electronic configurations, why arsenic has a higher first ionisation energy than selenium. **(2 marks)**

Look at the Periodic Table on page 151 to help you answer Questions 3 and 4.

Exam skills 1

This exam-style question uses knowledge and skills you have already revised. Have a look at pages 1–5 for a reminder about **relative atomic mass, mass spectrometry, electronic configuration ionisation** and **energy.**

Worked example

(a) Complete the electronic configurations for:

  (i) the chlorine atom

   $1s^2\, 2s^2\, 2p^6\, 3s^2\, 3p^5$ **(1 mark)**

  (ii) the chromium atom

   $1s^2\, 2s^2\, 2p^6\, 3s^2\, 3p^6\, 3d^5\, 4s^1$ **(1 mark)**

(b) A sample of chromium is made up of four isotopes. The data below were taken from a mass spectrum of this sample.

Mass/charge ratio	50	52	53	54
% abundance	4.35	83.79	9.50	2.36

Calculate the relative atomic mass of the sample, giving your answer to **two** decimal places. **(2 marks)**

$$\frac{(50 \times 4.35) + (52 \times 83.79) + (53 \times 9.50) + (54 \times 2.36)}{100}$$

$$= \frac{(217.5 + 4357.08 + 503.5 + 127.44)}{100}$$

$$= \frac{5205.52}{100} = 52.06$$

(c) Write an equation, including state symbols, to show the reaction that happens when the first ionisation energy of chlorine is measured.

  (1 mark)

$Cl(g) \rightarrow Cl^+(g) + e^-$

(d) Explain the general trend in first ionisation energies for the elements boron to neon (B to Ne) in Period 2. **(3 marks)**

First ionisation energy generally increases. This is because there is an increase in nuclear charge but there is similar shielding. The force of attraction between the nucleus and the outer electrons increases, so more energy is needed to remove an electron.

(e) There is a similar trend in first ionisation energies for the elements gallium to krypton (Ga to Kr) in Period 4. State how selenium deviates from this trend and explain your answer. **(3 marks)**

The first ionisation energy of selenium is lower than that of arsenic, the element before it.

The electron removed is one of two paired electrons in the 4p-sub-shell. Repulsion between these electrons means that less energy is needed to remove one of them.

Command word: Complete

If a question asks you to **complete** something, it means you need to finish a table or a diagram. You should follow the style used in the question.

Command word: Calculate

If a question asks you to **calculate** something, it means you need to:

- obtain a numerical answer
- show relevant working
- include the unit if the answer has one.

Maths skills If you are asked to give your answer to 3 significant figures, for example, you could round your intermediate values to 4 significant figures. Otherwise, only round your final answer.

Relative atomic mass has no units, but molar mass has units of $g\,mol^{-1}$.

You could write instead:

  $Cl(g) + e^- \rightarrow Cl^+(g) + 2e^-$

Do not confuse ionisation of chlorine molecules in the mass spectrometer (to produce Cl_2^+) with ionisation of its atoms.

You need to know the reasons for the trends in melting and boiling temperatures, and ionisation energies, in Periods 2 and 3.

Make sure you get the trend correct.
You could mention the increase in the number of protons instead of nuclear charge.

This question asks you to apply your knowledge of Periods 2 and 3 to a different period.

You could give your answer in terms of a 4p-orbital, rather than a sub-shell.

You could sketch an 'arrows in boxes' diagram to make your answer clear, e.g.

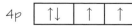

4p | ↑↓ | ↑ | ↑ |

Ions

An **ion** is a charged particle formed when an atom or group of atoms has lost, or gained, one or more electrons.

Positively charged ions

These form when electrons are lost.

For example:

$$Na \rightarrow Na^+ + e^- \qquad \text{and} \qquad Al \rightarrow Al^{3+} + 3e^-$$

During electrolysis, positively charged ions are attracted to the cathode (the negatively charged electrode), so they are called **cations**.

These atoms commonly form cations:
- metals
- hydrogen, forming the hydrogen ion H^+.

There are also positively charged compound ions such as ammonium, NH_4^+.

Negatively charged ions

These form when electrons are gained.

For example:

$$Cl + e^- \rightarrow Cl^- \qquad \text{and} \qquad O + 2e^- \rightarrow O^{2-}$$

During electrolysis, negatively charged ions are attracted to the anode (the positively charged electrode), so they are called **anions**.

These atoms commonly form anions:
- non-metals
- hydrogen, forming the hydride ion H^-.

There are also negatively charged compound ions such as nitrate, NO_3^-.

Configurations of s- and p-block ions

You must be able to predict the electronic configurations of ions formed from s- or p-block atoms. For example, magnesium and fluorine:

- Mg atom is 2.8.2 and Mg^{2+} ion is 2.8
- F atom is 2.7 and F^- ion is 2.8.

Mg^{2+} ions and F^- ions are **isoelectronic** – they have the same electronic configuration:

1s	2s	$2p_x$	$2p_y$	$2p_z$
↑↓	↑↓	↑↓	↑↓	↑↓

Evidence for ions

The migration of ions during electrolysis provides evidence for the existence of ions.

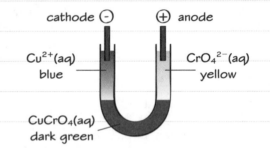

cathode ⊖   ⊕ anode

$Cu^{2+}(aq)$
blue

$CrO_4^{2-}(aq)$
yellow

$CuCrO_4(aq)$
dark green

Draw a diagram, using dots and crosses, to show the bonding in sodium oxide. Include all the electrons in each species and the charges present. **(3 marks)**

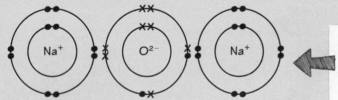

Be careful! You might be asked instead to show just the outer electrons, rather than all the electrons.

Do not let the circles touch or overlap because this incorrectly implies covalent bonding.

One mark each for the correct:
- number of electrons on each ion
- charges and symbols for each ion
- ratio or number of ions.

Note that Na^+ and O^{2-} ions are isoelectronic.

Now try this

1 A drop of green nickel(II) sulfate solution is added to the centre of a piece of moist filter paper, and the two ends are clipped to a DC power supply. Which answer (A, B, C or D) describes what is seen after a few minutes? **(1 mark)**

	Near the negative terminal	Near the positive terminal
A	green colour	no visible change
B	no visible change	green colour
C	yellow colour	blue colour
D	blue colour	yellow colour

2 State the electronic configurations of the following ions.
(a) K^+ (b) Ca^{2+} (c) S^{2-} (d) N^{3-} **(4 marks)**

3 Draw a dot and cross diagram to show the electronic structure of the compound calcium chloride. Only the outer electrons need be shown, but include the charges present. **(3 marks)**

Ionic bonds

Ionic bonding is the strong electrostatic force of attraction between oppositely charged ions.

Giant ionic lattice

The structure of solid ionic compounds comprises oppositely charged ions arranged in a highly ordered way – a **giant ionic lattice**.

repeated many times   ordered arrangement

type of bonding

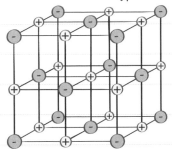

○ = Na^+
● = Cl^-

There are forces of **attraction** between ions with opposite charges (the ionic bonds).

There are also forces of **repulsion**:
- between ions with the same charges
- between nuclei, and between electrons.

Representing a lattice simply

Draw eight linked circles in a cube shape, and label one +.

Then add – signs and + signs as shown.

Group trends in ionic radius

Ionic radius increases down a group:
- the ions have more occupied shells.

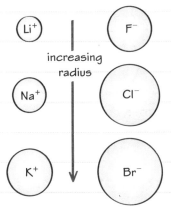

increasing radius

Isoelectronic ions

Isoelectronic ions have the same electronic configuration and number of electrons.

These ions all have ten electrons, $1s^2\,2s^2\,2p^6$:

N^{3-} O^{2-} F^- Na^+ Mg^{2+} Al^{3+}
  7   8   9   11   12   13

As the atomic number increases:
- the nuclear charge increases
- the attraction between the nucleus and electrons increases
- the ionic radius decreases.

The ionic bond between F^- ions and Mg^{2+} ions is stronger than that between F^- ions and Li^+ ions. Explain why this is so.   **(2 marks)**

The sizes of lithium ions and magnesium ions are similar, but magnesium ions have a higher charge. This means that magnesium ions have a higher charge density and are more strongly attracted to fluoride ions.

Ionic charge and ionic radius affect the strength of ionic bonding. The strength increases with:
- increase in ionic charge
- decrease in ionic radius.

Charge density is a measure of the charge/size ratio. In general, ions with a high charge density form stronger bonds than those with a low charge density.

Now try this

1 (a) Which ion (S^{2-}, Cl^-, K^+ or Ca^{2+}) has the largest ionic radius?   **(1 mark)**
  (b) Explain your answer.   **(3 marks)**

2 (a) Describe the trend in ionic radius in Group 2.   **(1 mark)**
  (b) Explain this trend.   **(1 mark)**

3 The ionic bond between Mg^{2+} ions and O^{2-} ions is stronger than that between Mg^{2+} ions and F^- ions. Explain why this is so.   **(3 marks)**

Covalent bonds

A **covalent bond** is the strong electrostatic attraction between two nuclei and the shared pair of electrons between them.

Representing a covalent bond

A covalent bond can be shown as:
- a straight line, e.g. H–H
- a dot and a cross in overlapping circles, e.g.

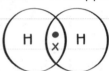

- a dot and cross between two symbols without the circles, e.g.

$$H \overset{\bullet}{\underset{\times}{}} H$$

Dot and cross diagrams

Show each electron in the outer shell as a dot or as a cross.

It does not matter which atom's electrons you show as dots, but you should show the electrons of the other atom as crosses.

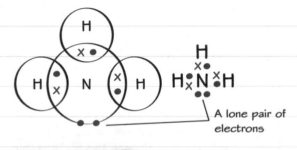

A lone pair of electrons

Overlapping orbitals

Two orbitals may overlap in different ways forming:
- sigma bonds, σ
- pi bonds, π

Bond	Orbitals	Overlap	Diagram
σ	s	end on	
σ	p	end on	
π	p	sideways	

Page 51 covers this for the alkene C=C bond.

Dative covalent bonds

A **dative covalent bond** forms when an orbital with a lone pair of electrons in one atom overlaps with a vacant orbital in another atom.

It is represented by an arrow pointing:
- from the atom that provides the electron pair
- towards the atom with the vacant orbital.

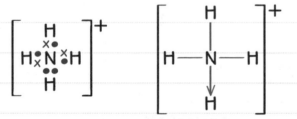

Worked example

Ammonia reacts with boron trifluoride to produce a compound whose bonding can be represented as $H_3N \rightarrow BF_3$. Draw a dot and cross diagram to show the bonding in this compound.   **(2 marks)**

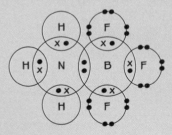

The N atom is providing the pair of electrons for the dative covalent bond, and these two electrons have been shown as dots.

This means that:
- electrons in H and B are shown as crosses
- electrons in F are shown as dots.

One mark is for correctly showing the shared pairs of electrons, and one mark for the lone pairs of electrons.

Now try this

1 Define the term covalent bond.   **(2 marks)**

2 Draw dot and cross diagrams for:
   (a) methane, CH_4 **(1 mark)**
   (b) water, H_2O **(1 mark)**
   (c) boron trichloride, BCl_3 **(1 mark)**

3 The bonding in the hydroxonium ion, H_3O^+, can be represented as $[H_2O \rightarrow H]^+$.
   (a) What type of bond does the arrow represent?   **(1 mark)**
   (b) Explain how this bond arises.   **(2 marks)**

Covalent bond strength

In general for covalent bonds of a similar nature, the shorter the bond, the stronger it is.

Two definitions

Bond length is the distance between the nuclei of two atoms that are covalently bonded together.

Bond lengths are measured in nanometres, nm ($1\,nm = 10^{-9}\,m$).

 Maths skills You must use appropriate units and be able to convert between units.

Bond strength is given by the bond enthalpy for a particular covalent bond:

- **Bond enthalpy** is the enthalpy change when one mole of a bond in the gaseous state is broken.
- Bond enthalpies are measured in $kJ\,mol^{-1}$.

← See page 65 for information about enthalpy changes.

Multiple bonds

A **multiple bond** arises when:
- two electron pairs are shared (double bonds), or
- three electron pairs are shared (triple bonds).

double bond in oxygen, O=O   triple bond in nitrogen, N≡N

Worked example

The bonding in dinitrogen monoxide, N_2O, can be represented as: N≡N→O. Complete the dot and cross diagram below.   **(3 marks)**

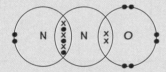

The right hand N atom provides the pair of electrons for the dative covalent bond.

These are shown as crosses.

Electrons in O and the other N are shown as dots. One mark each for:
- the six N≡N electrons
- the two N→O electrons
- the four lone pairs of electrons.

Single bonds, different atoms

For Cl_2, Br_2 and I_2, as the bond length increases the halogen–halogen bond becomes weaker.

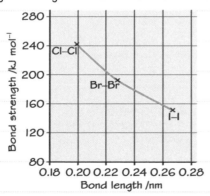

The F–F bond is shorter and weaker than the Cl–Cl bond because its lone pairs of electrons are close enough to produce a lot of repulsion.

Same atoms, different bonds

For a carbon–carbon bond, as the number of bonds decreases:
- the bond length increases
- the bond strength decreases.

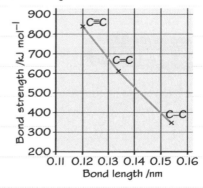

The electrostatic attraction between the two nuclei and the shared electrons is decreased.

Now try this

1 (a) Describe the relationship you expect to see between the bond length and bond strength for N–N, N=N and N≡N.   **(2 marks)**
   (b) Explain your answer to part (a).   **(1 mark)**

2 N=N and O=O bonds are almost identical in length but the O=O bond is about 21% stronger. Suggest a reason for this.   **(1 mark)**

3 Draw a dot and cross diagram to show the bonding in carbon dioxide, O=C=O.   **(2 marks)**

Shapes of molecules and ions

Valence shell electron pair repulsion (VSEPR) theory lets you predict shapes of molecules and ions.

Predicting a shape

Use a dot and cross diagram to find the number of pairs of electrons around the central atom.
This gives you the basic shape (see table below).

Multiple bonds

Treat these as single bonds.

For example, $O=C=O$ and $H-C\equiv C-H$ are linear molecules with a bond angle of 180°.

> **Maths skills** Make sure you can visualise the angles and shapes in regular 2D and 3D shapes and can draw them.

The effect of lone pairs

The arrangement of electron pairs around the central atom keeps repulsion to a minimum.

lone pair–lone pair	most repulsion
lone pair–bond pair	↓
bond pair–bond pair	least repulsion

Each lone pair of electrons reduces the bond angle by about 2.5°. For example, NH_3 is trigonal pyramidal, bond angle 107° (compared with 109.5° if it were tetrahedral).

Shapes and angles

Bond pairs	Lone pairs	Shape	Example		Bond pairs	Lone pairs	Shape	Example	
2	0	linear	Cl—Be—Cl 180°	Beryllium is in Group 2.	5	0	trigonal bipyramidal	P with Cl's, 90°, 120°	A dashed line shows a bond going into the plane of the paper.
3	0	trigonal planar	Cl—B with Cl, Cl 120°	Boron is in Group 3.	6	0	octahedral	S with F's, 90°	A wedge shows a bond coming out of the plane of the paper.
4	0	tetrahedral	C with H's 109.5°	An ordinary line shows a bond in the plane of the paper.					

Draw the shape of a water molecule, including its bond angle.   **(1 mark)**

There are two bonding pairs and two lone pairs around the O atom, so the basic shape is tetrahedral.

Repulsion by the two lone pairs reduces the bond angle by about 5°, producing a V-shaped (or bent line) molecule.

Now try this

1 Predict the bond angles in the following molecules and ions.
   (a) $CHCl_3$ (b) NH_4^+ (c) NH_2^- (d) BeF_2 (e) PF_5
                                                              (5 marks)

   Carbon is the central atom in $CHCl_3$.

2 Predict the bond angles in BH_3 and PH_3. Explain the difference between them.   **(3 marks)**

3 Sulfur trioxide, SO_3, has three triple bonds around the central atom and no lone pairs. Name the shape of the SO_3 molecule and suggest its bond angle.   **(2 marks)**

Electronegativity and bond polarity

Electronegativity is the ability of an atom to attract the bonding electrons in a covalent bond.

Trends in electronegativity

In general, electronegativity increases:
• up a group
• across a period.

H 2.1							He
Li 1.0	Be 1.5	B 2.0	C 2.5	N 3.0	O 3.5	F 4.0	Ne
Na 0.9	Mg 1.2	Al 1.5	Si 1.8	P 2.1	S 2.5	Cl 3.0	Ar

Fluorine is the most electronegative element. It is given a value of 4.0 on the Pauling electronegativity scale.

Polar covalent bonds

A covalent bond is **polar** if the two bonded atoms have different electronegativities.

For example, the H–Cl bond is polar:

delta plus = partial positive charge ——— δ^+ δ^- ——— delta minus = partial negative charge

H——Cl

——— represents a dipole

H–H and Cl–Cl have **non-polar** bonds:
• the atoms in their molecules have identical electronegativities.

Ionic versus covalent

Covalent and ionic bonding are the two extremes of a continuum of bonding types.

Polar bonds can be described as covalent bonds with a degree of **ionic character**.

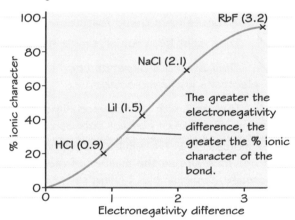

The greater the electronegativity difference, the greater the % ionic character of the bond.

Polar bonds and polar molecules

A **polar bond** has a permanent **dipole**:
• the less electronegative atom has a δ^+ charge
• the more electronegative atom has a δ^- charge.

A **polar molecule** must have:
• one or more polar bonds
• unbalanced dipoles because of their partial charges or the shape of the molecule.

Water molecules are polar.

H_2O molecules have two identical dipoles in a V-shape.

Carbon dioxide molecules are not polar.

CO_2 molecules have two identical dipoles in a straight line.

Worked example

Explain why a stream of $CHCl_3$ is deflected by a charged rod but a stream of CCl_4, is not. **(2 marks)**

$CHCl_3$ contains polar bonds and is also a polar molecule, so it is affected by an electric field. CCl_4 contains polar bonds but is not a polar molecule, so it is unaffected.

Both molecules have a tetrahedral shape with four polar bonds, but the dipoles in CCl_4 are identical and cancel each other out.

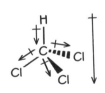
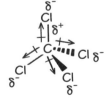

Now try this

1 (a) Define the term electronegativity. **(1 mark)**

(b) Describe the trends in electronegativity in the Periodic Table. **(2 marks)**

2 Explain why gaseous hydrogen halides (HF, HCl, HBr and HI) are mainly covalent but with an increasing ionic character as you go up the group. **(2 marks)**

3 Which of these molecules contains polar bonds but is not a polar molecule? **(1 mark)**
Br_2, $CHBr_3$, HBr, CBr_4

London forces

London forces are a type of **intermolecular force**, forces that act between molecules.

Instantaneous dipoles

Permanent dipoles exist because of a difference in electronegativity between two covalently bonded atoms.

Instantaneous dipoles do not need differences in electronegativity – they can exist in non-polar molecules.

This is because electron density in a molecule fluctuates constantly.

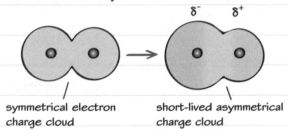

symmetrical electron   short-lived asymmetrical
charge cloud   charge cloud

Induced dipoles

An **induced dipole** is caused when a molecule comes close to a molecule with a permanent dipole or an instantaneous dipole. Electrons are:
• repelled by δ⁻ charges
• attracted to δ⁺ charges.

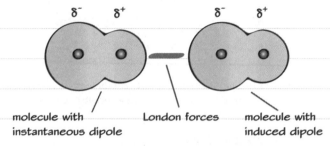

molecule with   London forces   molecule with
instantaneous dipole   induced dipole

London forces are attractive forces between instantaneous dipoles and induced dipoles.

Noble gases are monatomic

The elements in Group 0 exist as single atoms.
London forces are the only forces of attraction between them.

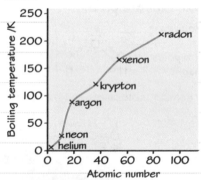

Features of London forces

London forces have these features:

☑   They exist between all simple molecules.

☑   Their strength depends upon the number of electrons in the molecule.

☑   Their strength also depends upon the number of points of contact between molecules.

For molecules with a similar shape, London forces increase as the number of electrons increase. For example, boiling temperature increases as:
• the A_r increases in noble gases (see graph on the left)
• the M_r increases in alkanes and halogens.

Butane boils at 273 K and 2-methylpropane at 261 K, but both contain the same number of electrons. Explain the difference. **(3 marks)**

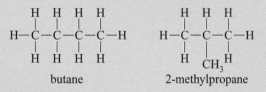

butane   2-methylpropane

2-methylpropane has a branch but butane does not. The branch reduces the number of points of contact between molecules, so London forces between 2-methylpropane molecules are weaker. These are more easily overcome than the forces between butane molecules, so 2-methylpropane has a lower boiling temperature.

1 Describe how London forces arise.   **(3 marks)**

2 Explain, in terms of the intermolecular forces present, why chlorine is a gas at room temperature, bromine is a liquid and iodine is a gas.   **(3 marks)**

3 These molecules all have the same number of electrons. Explain which one will have the highest boiling temperature.   **(1 mark)**

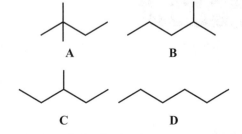

A   B

C   D

Permanent dipoles and hydrogen bonds

Some molecules can form intermolecular forces in addition to London forces.

Forces between permanent dipoles

Molecules with permanent dipoles have intermolecular forces of attraction and repulsion depending on their orientation.

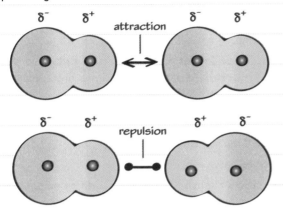

This means the overall force of attraction is usually less than the London forces present.

Hydrogen bonds

Hydrogen bonds are intermolecular forces.

In general they can form if:

✓   the compound contains a hydrogen atom covalently bonded to an atom of N, O or F

✓   there is a lone pair of electrons on an atom of N, O or F.

Hydrogen bonds are important in water.

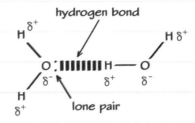

The bond angle between the three atoms involved in a hydrogen bond is often 180° (as shown here).

Trends in hydrogen halides

The main intermolecular forces in HCl, HBr and HI are London forces:

• their boiling temperatures increase as the number of electrons per molecule increases.

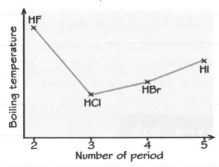

HF has the highest boiling temperature, even though its London forces are the weakest.

This is because it also has hydrogen bonding between molecules.

Hydrogen bonding and water

Water has anomalous properties:
• relatively high melting and boiling temperatures for a molecule with its number of electrons
• its solid (ice) is less dense than its liquid.

Each water molecule can form two hydrogen bonds on average, increasing the amount of energy needed to melt or boil water.

Ice forms open rings of six water molecules joined by hydrogen bonds.

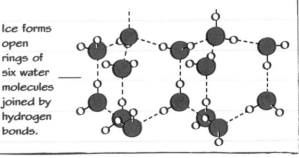

Worked example

Ethanol is a liquid at room temperature and propane is a gas, but both have similar numbers of electrons. Explain the difference.   **(3 marks)**

Both ethanol and propane have London forces, but ethanol contains a hydroxyl group –OH. This allows it to form hydrogen bonds (to other ethanol molecules). More energy is needed to separate ethanol molecules than to separate propane molecules.

Alcohols have low volatility compared to alkanes with a similar number of electrons.

Now try this

1 Methane, CH_4, and ammonia, NH_3, have similar numbers of electrons, yet one has a much higher boiling temperature than the other. Identify the compound with the higher boiling temperature, and explain your answer.   **(2 marks)**

2 (a) Explain why hydrogen fluoride can form hydrogen bonds but hydrogen bromide cannot do so.   **(2 marks)**

   (b) Draw a labelled diagram to show hydrogen bonding between HF molecules.   **(2 marks)**

Choosing a solvent

A **solvent** is a substance in which another substance can dissolve, forming a solution.

Conditions for a substance to dissolve

A substance will dissolve in a given solvent if:

- the solute particles can be separated from one another
- the separated solute particles become surrounded by solvent particles and
- the solute–solvent forces are greater than the solute–solute and solvent–solvent forces.

Energy is released when bonds form.

Some definitions

A **solute** is the substance that dissolves in the solvent. For example in salty water, salt is the solute and water is the solvent.

A solute is **soluble** if it can dissolve in a given solvent.

A **solution** is the mixture formed between a solute and its solvent. Solutions are clear and can be coloured or colourless.

Water and ionic compounds

Ionic compounds are often soluble in water.

Water molecules surround individual ions and form strong electrostatic attractions with them. This is **hydration**.

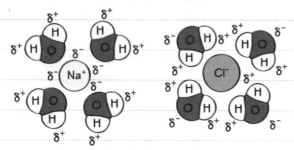

For example, when sodium chloride dissolves, **ion–dipole interactions** form between:

- Na^+ ions and the δ^- of water molecules
- Cl^- ions and the δ^+ of water molecules.

Water and alcohols

Alcohols have a hydroxyl group –OH.

This allows them to form hydrogen bonds.

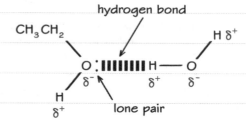

Methanol CH_3OH, ethanol C_2H_5OH and propanol C_3H_7OH are completely soluble in water.

As the number of carbon atoms in an alcohol molecule increases:

- London forces become more important
- the alcohol becomes less soluble.

Worked example

In terms of hydrogen bonding, suggest why chlorobutane C_4H_9Cl is less soluble in water than it is in ethanol. **(2 marks)**

Water forms more hydrogen bonds per molecule than ethanol does. More energy would be needed to break the hydrogen bonds in water than in ethanol, so chlorobutane is less soluble in water than it is in ethanol.

Water molecules form two hydrogen bonds on average while ethanol molecules only form one.

Chlorobutane molecules are polar but they cannot form hydrogen bonds.

Be careful! Polar substances usually dissolve in polar solvents, and non-polar substances usually dissolve in non-polar solvents such as hexane.

Now try this

1 Why is potassium chloride more soluble in water than it is in hexane? Choose A, B, C or D.
   - ☐ **A** Potassium ions and chloride ions form hydrogen bonds with water.
   - ☐ **B** Hexane molecules are too large to fit between ions in the giant ionic lattice.
   - ☐ **C** Hexane molecules have weaker intermolecular forces than water molecules.
   - ☐ **D** Energy is released when the ions in potassium chloride are hydrated. **(1 mark)**

2 Propanone CH_3COCH_3 is a widely used liquid solvent.
   (a) Explain why alkanes dissolve in propanone. **(1 mark)**
   (b) Explain why water also dissolves in propanone. **(1 mark)**

16

Giant lattices

Giant lattices are present in solid metals, ionic solids and covalently bonded solids like diamond.

Metals and metallic bonding

Solid metals have giant metallic lattices in which positive metal ions are:

✓ closely packed in a regular lattice
✓ surrounded by delocalised electrons.

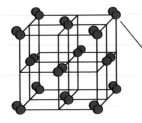

delocalised electron

metal cation

Metallic bonding is the strong electrostatic force of attraction between metal ions and the delocalised electrons.

Ions and ionic bonding

Solid ionic compounds have **giant ionic lattices** with oppositely charged ions that are:

✓ closely packed in a regular lattice.

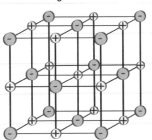

Ionic bonding is the strong electrostatic force of attraction between oppositely charged ions.

For example, between Na^+ and Cl^- ions (above).

Structure of iodine

Iodine exists as simple **diatomic** molecules, I_2.

In solid iodine, these molecules are:

✓ arranged in a simple molecular lattice
✓ attracted to each other by London forces.

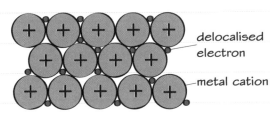

There are similarities in structure between an iodine crystal and a sodium chloride crystal.

Diamond and graphite

Diamond and graphite are forms of carbon.

They exist as **giant covalent lattices**.

Diamond	Graphite
each C atom joined to 4 others by σ covalent bonds	each C atom joined to 3 others by σ covalent bonds
tetrahedral arrangement of atoms	interlocked hexagonal rings of atoms in layers

Graphite has London forces between its layers, with delocalised electrons above and below each plane of atoms.

Worked example

Graphene is a form of pure carbon. It exists in sheets just one atom thick. What type of crystal structure does graphene have?

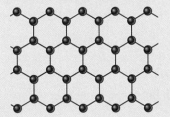

(1 mark)

Graphene has a giant covalent lattice structure.

Notice that graphene has a structure identical to a single layer of graphite:
• each C atom is joined to 3 others by σ bonds
• the C atoms are arranged in hexagonal rings.

It is a good electrical conductor, like graphite.

Now try this

1 Describe the structure and bonding present in:
   (a) sodium   **(3 marks)**
   (b) sodium chloride   **(3 marks)**
   (c) iodine   **(3 marks)**

2 The structure and bonding in silicon(IV) oxide are similar to those found in diamond.

   Describe the structure and bonding present in silicon(IV) oxide.

   (3 marks)

Structure and properties

The physical properties of a substance may be explained by its structure, particles and bonding.

Typical properties

Structure	giant metallic	giant ionic	giant covalent	molecular
Particles	metal ions delocalised electrons	positive ions negative ions	atoms	simple molecules
Bonding	metallic	ionic	covalent	covalent
Intermolecular forces	✗	✗	✗	✓
Melting and boiling temperatures	fairly high to high	fairly high to high	high to very high	generally low
Good electrical conductor?	✓ solid ✓ liquid	✗ solid ✓ liquid ✓ in solution	✗	✗
Soluble in water?	✗	✓	✗	✗

Metals like sodium react with water to produce soluble products.

Some are insoluble, such as AgCl, AgBr, AgI and $BaSO_4$.

Graphite has delocalised electrons and does conduct electricity.

May be soluble if they can form hydrogen bonds (e.g. glucose) or they react with water (e.g. chlorine).

Worked example

The structure and bonding in the hexagonal form of boron nitride is similar to that of graphite. Predict, with reasons, its melting temperature and electrical conductivity. **(3 marks)**

It should have a high melting temperature because of its giant covalent lattice structure with many strong covalent bonds that must be broken. It should conduct electricity when solid because it will have delocalised electrons.

Be careful!

You should also be able to predict the solubility in water of different substances and to predict the type of structure present from given data.

The answer links each property to the relevant type of particle, type of bonding or type of structure.

Boron nitride is also insoluble in water as it is unable to form hydrogen bonds or to react with water.

Now try this

The table below gives some properties of four substances (**A**, **B**, **C** and **D**).

Use the data to predict the type of structure and bonding present in each one, explaining your answers fully. **(8 marks)**

Substance	Melting temperature/°C	Boiling temperature/°C	Ability to conduct electricity	Solubility in water
A	1085	2562	good when solid or molten	insoluble
B	2072	2977	not when solid good when molten	insoluble
C	–7.2	59	not when solid or molten	slightly soluble
D	1680	2230	not when solid or molten	insoluble

Exam skills 2

This exam-style question uses knowledge and skills you have already revised. Have a look at pages 9, 12, 14, 15 and 17 for a reminder about **ionic bonds**, **shapes of molecules and ions**, **London forces**, **hydrogen bonds** and **giant lattices**.

Worked example

(a) Explain why ionic compounds have relatively high melting temperatures. **(2 marks)**

They have strong electrostatic forces of attraction between their ions. The ions are held in giant ionic lattices so there are many ionic bonds.

(b) Explain why the boiling temperature of aluminium is higher than the boiling temperature of magnesium. **(3 marks)**

Aluminium ions have a greater charge than magnesium ions. Aluminium ions are also smaller than magnesium ions.
This means that aluminium ions have a greater attraction for the 'sea' of delocalised electrons, so more energy is needed to overcome these forces.

(c) (i) Draw and name the shape of a silicon tetrachloride, $SiCl_4$, molecule. Predict its bond angle. **(2 marks)**

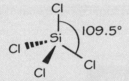

Tetrahedral.

(ii) Draw and name the shape of a phosphine, PH_3, molecule. Predict its bond angle. **(2 marks)**

Trigonal pyramidal.

(d) Explain why ammonia has a higher boiling temperature than methane. **(3 marks)**

Both substances exist as simple molecules with instantaneous dipole-induced dipole forces between them. However, ammonia molecules also have hydrogen bonds.
These are stronger and need more energy to overcome them.

Command word: Explain

If a question asks you to **explain** something, it means you need to give reasons or justifications for your answer.

 You could also mention that a large amount of energy is needed to break the bonds.

Make sure you refer to the correct type of bonding and not to ionic bonding, or to molecules and intermolecular forces.

You could mention that Al^{3+} ions are smaller than Mg^{2+} ions instead of the first two sentences, as their charges are shown. Aluminium atoms also contribute more delocalised electrons.

Command word: Draw

If a question asks you to **draw** something, it means you need to produce a diagram freehand, or using a ruler.

Command word: Predict

If a question asks you to **predict** something, it means you need to give an expected result.
Here, you are using the electron-pair repulsion theory to predict bond angles.

The four pairs of electrons around the central P atom (three bonding pairs and one lone pair) repel to the maximum extent.

Lone pairs repel more than bonding pairs, so the bond angle is less than the tetrahedral angle ($109.5°$).

 This question asks you to apply your knowledge of intermolecular forces.

Instantaneous dipole-induced dipole forces are also called London forces.

Remember that the presence of hydrogen atoms alone does not lead to hydrogen bonding, otherwise they would also be present in methane.

Oxidation numbers

The **oxidation number** of an element represents the number of electrons lost or gained by an atom of that element in a compound or ion.

Rules for calculating oxidation numbers

Here are some basic rules:

- The oxidation number of an uncombined element is 0.
- The total of all the oxidation numbers in a neutral molecule is 0.
- The oxidation number of a monatomic ion is the same as its charge.
- The total of all the oxidation numbers in an ion is the same as its charge.
- Oxidation states are shown as sign then number, e.g. +2 rather than 2+.

Additional rules

Element(s)	Rule
fluorine	−1 always
Group 7 except fluorine	−1 unless bonded to O or N, or to a more electronegative Group 7 element
oxygen	−2 unless in a peroxide (when it is −1)
hydrogen	+1 unless bonded to Na, Al or B (when it is −1)
Group 1	+1
Group 2	+2

Four example calculations

1. Al^{3+} ——————— +3 because it is a monatomic ion

2. SO_2
   -2
   +4 because there two oxygen atoms at −2 each and the total must be 0

3. HSO_3^-
   $+1$ -2
   +4 because there are three oxygen atoms at −2 each and the total must be −1

4. $NaClO$
   $+1$ -2
   +1 because the total must be 0 (notice that Cl is not −1 here)

Systematic names and formulae

When an element can have more than one oxidation state, its oxidation state is shown in its name:

- Cu_2O is copper(I) oxide
- CuO is copper(II) oxide

Roman numeral 2 in brackets, with no space between the element's name and the opening bracket.

If an element is bonded to oxygen, its name is often given the ending 'ate':

- $KMnO_4$ is potassium manganate(VII)
- $Cr_2O_7^{2-}$ is the dichromate(VI) ion

Roman numeral 6

Worked example

(a) Calculate the oxidation number of the stated element in each species. **(4 marks)**

(i) N in NH_3

−3

(ii) Cl in ClO_3^-

+5

(iii) S in H_2SO_4

+6

(iv) H in $LiAlH_4$

−1

This is lithium aluminium hydride. The oxidation number of hydrogen is usually +1, but when hydrogen is bonded to aluminium it is −1.

(b) Write the formula for chromium(III) nitrate(V). **(1 mark)**

$Cr(NO_3)_3$

The chromium(III) ion is Cr^{3+}.
The nitrate(V) ion is NO_3^-.

Now try this

1 Calculate the oxidation number of chlorine in the following species.
   (a) Cl_2 (b) Cl^- (c) $CaCl_2$ (d) ClO_2 (e) ClO_4^-
   (5 marks)

2 Calculate the oxidation number of nitrogen in the following species.
   (a) N_2H_4 (b) NCl_3 (c) N_2O_3 (d) NH_4^+ (e) HNO_2
   (5 marks)

3 State the systematic names for the following species.
   (a) PCl_5 (b) Na_2CrO_4 (c) VO_2^- **(3 marks)**

4 Deduce the formulae for the following compounds.
   (a) iron(III) chloride
   (b) lead(IV) oxide
   (c) ammonium vanadate(V) **(3 marks)**

The vanadate(V) ion has a charge of −1.

Redox reactions

Redox reactions involve both reduction and oxidation. They can be described in terms of transfer of electrons or changes in oxidation number.

Oxidation

Oxidation can be described in three ways:

- gain of oxygen
- loss of electrons
- increase in oxidation number.

In general, metals form positive ions by:

- loss of electrons
- increase in oxidation number.

For example, sodium atoms form sodium ions.

$$Na \rightarrow Na^+ + e^-$$

oxidation number 0   oxidation number +1

Reduction

Reduction can be described in three ways:

- loss of oxygen
- gain of electrons
- decrease in oxidation number.

In general, non-metals form negative ions by:

- gain of electrons
- decrease in oxidation number.

For example, chlorine atoms form chloride ions.

$$Cl + e^- \rightarrow Cl^-$$

oxidation number 0   oxidation number −1

Identifying oxidation and reduction

Sodium reacts with chlorine:

$$2Na + Cl_2 \rightarrow 2NaCl$$

Each sodium atom transfers an electron to a chlorine atom.

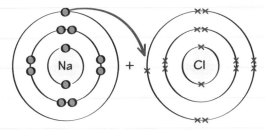

In the example above:

- sodium acts as a reducing agent because it transfers electrons to chlorine
- chlorine acts as an oxidising agent because it receives electrons from sodium.

Oxidising and reducing agents

Oxidising agents gain electrons – they are electron acceptors.

Reducing agents lose electrons – they are electron donors.

Sodium is oxidised and chlorine is reduced.

A useful memory aid is OIL-RIG:

Oxidation Is Loss of electrons

Reduction Is Gain of electrons.

Be careful! Some reactions look like they are redox reactions but they are not.
For example:

$$NaCl + H_2SO_4 \rightarrow NaHSO_4 + HCl$$

Concentrated sulfuric acid can act as an oxidising agent, but there is no change in any of the oxidation numbers here.

Worked example

Sodium chlorate(I) forms sodium chloride and sodium chlorate(V) when it is heated:

$$3NaClO(aq) \rightarrow 2NaCl(aq) + NaClO_3(aq)$$

Identify the type of reaction involved and explain your answer. **(3 marks)**

This is a disproportionation reaction, because chlorate(I) ions are both reduced to chloride ions and oxidised to chlorate(V) ions.

Look at the oxidation numbers to spot disproportionation. Here chlorine is +1 in NaClO, −1 in NaCl and +5 in $NaClO_3$.

- +1 → −1 is reduction.
- +1 → +5 is oxidation.

Now try this

One of the reactions that happens in a car's catalytic converter is:

$$2CO(g) + 2NO(g) \rightarrow 2CO_2(g) + N_2(g)$$

Explain, in terms of oxidation numbers, why this is a redox reaction. **(4 marks)**

Ionic half-equations

Reduction and oxidation happen simultaneously in redox reactions, but the two processes can be shown separately as **ionic half-equations**.

Simple half-equations

A redox reaction happens when zinc powder is added to blue copper(II) sulfate solution.

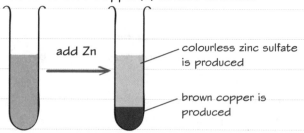

add Zn

colourless zinc sulfate is produced

brown copper is produced

$$Zn(s) + CuSO_4(aq) \rightarrow ZnSO_4(aq) + Cu(s)$$

This can be split into two ionic half-equations:

$$Zn(s) \rightarrow Zn^{2+}(aq) + 2e^- \text{ oxidation}$$
$$Cu^{2+}(aq) + 2e^- \rightarrow Cu(s) \text{ reduction}$$

Spectator ions

Spectator ions are present in a reaction mixture but do not take part in the reaction.

In the example, SO_4^{2-} ions are spectator ions.

The two half-equations can be combined to form a full ionic equation without the spectator ions:

$$Zn(s) + Cu^{2+}(aq) \rightarrow Zn^{2+}(aq) + Cu(s)$$

See below for more about combining equations.

Oxidation half-equations

For example, oxidation of iron(II) ions.

• Write down what you know:

$$Fe^{2+} \rightarrow Fe^{3+}$$

make sure the atoms balance

• Add electrons to the right hand side

$$Fe^{2+} \rightarrow Fe^{3+} + e^-$$ — make sure the charges balance

Reduction half-equations

For example, reduction of chlorine.

• Write down what you know:

$$Cl_2 \rightarrow 2Cl^-$$

make sure the atoms balance

• Add electrons to the left hand side

$$Cl_2 + 2e^- \rightarrow 2Cl^-$$

make sure the charges balance

Combining half-equations

Make sure both half-equations have the same number of electrons before combining them.

$$Fe^{2+} \rightarrow Fe^{3+} + e^-$$ ——one electron

$$Cl_2 + 2e^- \rightarrow 2Cl^-$$

two electrons

Multiply one or both equations as needed.

$$2Fe^{2+} \rightarrow 2Fe^{3+} + 2e^-$$
$$Cl_2 + 2e^- \rightarrow 2Cl^-$$

Add both left hand sides together and both right hand sides together. Leave the electrons out.

$$2Fe^{2+} + Cl_2 \rightarrow 2Fe^{3+} + 2Cl^-$$

• Check that the numbers of atoms of each element are the same on both sides.
• Check that the total charges are the same on both sides.

Worked example

Manganate(VII) ions are reduced in acidic conditions to manganese(II) ions. Write an ionic half-equation for this change. **(1 mark)**

$$MnO_4^- + 8H^+ + 5e^- \rightarrow Mn^{2+} + 4H_2O$$

You can work this out in stages:

$$MnO_4^- \rightarrow Mn^{2+}$$
$$MnO_4^- \rightarrow Mn^{2+} + 4H_2O \text{ (water for the O atoms)}$$
$$MnO_4^- + 8H^+ \rightarrow Mn^{2+} + 4H_2O \text{ (balancing H)}$$
$$MnO_4^- + 8H^+ + 5e^- \rightarrow Mn^{2+} + 4H_2O$$
(balancing charges)

Now try this

1 Combine these pairs of ionic half-equations to make full ionic equations.
   (a) $Ag^+ + e^- \rightarrow Ag$ and $Cu \rightarrow Cu^{2+} + 2e^-$ **(1 mark)**
   (b) $MnO_4 + 8H^+ + 5e^- \rightarrow Mn^{2+} + 4H_2O$ and
   $Fe^{2+} \rightarrow Fe^{3+} + e^-$ **(1 mark)**

2 Write an ionic half-equation for the reduction of hydrogen ions to form hydrogen gas. **(1 mark)**
3 Balance this ionic half-equation:
   $$H_2O_2 \rightarrow O_2$$
   (1 mark)

You will need hydrogen ions and electrons.

Reactions of Group 2 elements

The elements in Group 2 include beryllium, magnesium, calcium, strontium, barium and radium. See the Periodic Table on page 151.

Reactivity increases down Group 2

Group 2 elements form positively charged ions in reactions with oxygen, chlorine or water:

$$M \rightarrow M^{2+} + 2e^-$$

Going down the group:
- the energy needed to form 2+ ions decreases
- the reactivity increases.

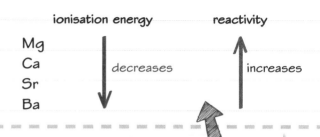

ionisation energy   reactivity

Mg
Ca   decreases   increases
Sr
Ba

Explain why the first ionisation energy of magnesium ($738 \, kJ \, mol^{-1}$) is greater than that of calcium ($590 \, kJ \, mol^{-1}$). **(2 marks)**

The atomic radius of magnesium is less than that of calcium, so the outer electrons are closer to the nucleus. There are fewer occupied shells in magnesium, so there is less shielding.

Going down Group 2:
- atomic radii increase
- first ionisation energies decrease.

These factors outweigh the increase in the nuclear charge going down Group 2.

Overall, the force of attraction between the outer electrons and the nucleus decreases.

Reactions with oxygen

The Group 2 elements (Mg to Ba) all react with oxygen, especially when heated.

In general,

$$2M(s) + O_2(g) \rightarrow 2MO(s)$$

Magnesium, for example, burns with a bright white flame to produce white magnesium oxide.

Barium is often stored under oil to keep it from reacting with oxygen and water vapour in air.

Reactions with chlorine

The Group 2 elements (Mg to Ba) all react with chlorine, especially when heated.

In general,

$$M(s) + Cl_2(g) \rightarrow MCl_2(s)$$

Barium chloride is toxic, like barium hydroxide and other soluble barium compounds.

Hot magnesium reacts vigorously with steam to produce magnesium oxide and hydrogen:

$$Mg(s) + H_2O(g) \rightarrow MgO(s) + H_2(g)$$

Reactions with water

The Group 2 elements (Mg to Ba) all react with water:

$$M(s) + 2H_2O(l) \rightarrow M(OH)_2(aq) + H_2(g)$$

bubbling

Mg
Ca   very slow
Sr
Ba   rapid

The rate of bubbling due to hydrogen increases going down the group.

Calcium hydroxide forms as a white precipitate, but $Ba(OH)_2$ forms a colourless solution.

Magnesium reacts very slowly with water but rapidly with steam. The hydrogen can be ignited.

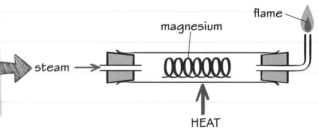

flame

magnesium

steam →

HEAT

---

1   Magnesium reacts with oxygen, and with steam, to produce magnesium oxide.
    (a)   Write equations, including state symbols, to represent these reactions.   **(4 marks)**
    (b)   Including an equation, describe the reaction of magnesium with water.   **(3 marks)**

2   State and explain the trend in reactivity down Group 2.   **(3 marks)**

3   Describe and explain a precaution taken when storing barium.   **(2 marks)**

Reactions of Group 2 compounds

You need to know the reactions of dilute acid with Group 2 oxides and hydroxides; the reactions of water with Group 2 oxides and trends in the solubility of the hydroxides and sulfates.

Hydroxide solubility increases down Group 2

Sparingly soluble, forms a white suspension called Milk of Magnesia, which is used as an antacid.

increasingly soluble (downward arrow)

$Mg(OH)_2$
$Ca(OH)_2$
$Sr(OH)_2$
$Ba(OH)_2$

Forms a solution called limewater, which is used to detect the presence of carbon dioxide.

Sulfate solubility decreases down Group 2

forms a colourless solution

increasingly soluble (upward arrow)

$MgSO_4$
$CaSO_4$
$SrSO_4$
$BaSO_4$

sparingly soluble

A non-toxic barium salt (because it is insoluble), used as a contrast medium in medical X-ray photographs.

Worked example

(a) Describe how to test a solution for the presence of sulfate ions. **(2 marks)**

Add a few drops of hydrochloric acid, followed by a few drops of barium chloride solution. A white precipitate of barium sulfate will form if sulfate ions are present.

(b) Write an ionic equation for the reaction seen. Include state symbols. **(2 marks)**

$$Ba^{2+}(aq) + SO_4^{2-}(aq) \rightarrow BaSO_4(s)$$

You need to write how to carry out the test, and what is seen if sulfate ions are present.

The acid prevents formation of ions such as sulfite, SO_3^{2-}, also giving a precipitate in the test.
You could use nitric acid and barium nitrate solution instead and get the same result.

The state symbol (s) shows that the reaction produces a solid. This is insoluble barium sulfate, which is why a white precipitate forms.

Reactions of oxides with water

The Group 2 oxides:

- are basic
- react with water in exothermic reactions to form hydroxides.

For example, $SrO(s) + H_2O(l) \rightarrow Sr(OH)_2(aq)$

In these reactions, the Group 2 ion acts as a spectator ion, so the equation can be shown as:

$$O^{2-} + H_2O \rightarrow 2OH^-$$

The presence of hydroxide ions makes the resulting solutions alkaline, with pH>7.

Going down Group 2:

- the hydroxides become increasingly soluble
- the pH of the resulting solution increases.

Hydroxides and oxides with acids

Group 2 hydroxides and oxides react with acids to form salts and water. These are:

- neutralisation reactions
- exothermic reactions.

For example:

- $MgO + H_2SO_4 \rightarrow MgSO_4 + H_2O$
- $CaO + 2HNO_3 \rightarrow Ca(NO_3)_2 + H_2O$
- $Sr(OH)_2 + H_2SO_4 \rightarrow SrSO_4 + 2H_2O$
- $Ba(OH)_2 + 2HCl \rightarrow BaCl_2 + 2H_2O$

Magnesium and water

A layer of insoluble magnesium hydroxide forms when magnesium is added to water. This is why bubbling is slow and stops after a while.

Now try this

Calcium oxide reacts vigorously with excess water to form a solution often called limewater.

(a) Write an equation, including state symbols, for the reaction. **(2 marks)**

(b) Describe what you would expect to see if universal indicator is added to limewater, giving a reason for your answer. **(2 marks)**

(c) Write an equation, including state symbols, for the reaction between limewater and hydrochloric acid. **(2 marks)**

Stability of carbonates and nitrates

Group 1 and 2 carbonates and nitrates may undergo **thermal decomposition**, reactions in which heat is used to break down a reactant into two or more products.

Carbonates

Many, but not all, Group 1 and 2 carbonates decompose to form metal oxides and carbon dioxide.

Li_2CO_3	decomposes	$BeCO_3$	decompose with increasing difficulty ↓
Na_2CO_3		$MgCO_3$	
K_2CO_3	do not decompose	$CaCO_3$	
Rb_2CO_3		$SrCO_3$	
Cs_2CO_3		$BaCO_3$	

In Group 1, only lithium carbonate decomposes at Bunsen burner temperatures:

$$Li_2CO_3(s) \rightarrow Li_2O(s) + CO_2(g)$$

Going down Group 2, the carbonates become more stable and need higher temperatures to decompose them. In general:

$$MCO_3(s) \rightarrow MO(s) + CO_2(g)$$

Nitrates

Group 1 and 2 nitrates decompose to form different products, depending on their stability.

$LiNO_3$	Li_2O NO_2 O_2	$Be(NO_3)_2$	
$NaNO_3$		$Mg(NO_3)_2$	
KNO_3	MNO_2	$Ca(NO_3)_2$	MO NO_2 O_2
$RbNO_3$	O_2	$Sr(NO_3)_2$	
$CsNO_3$		$Ba(NO_3)_2$	

Here are three example equations:

- $4LiNO_3(s) \rightarrow 2Li_2O(s) + 4NO_2(g) + O_2(g)$
- $2RbNO_3(s) \rightarrow 2RbNO_2(s) + O_2(g)$
- $2Mg(NO_3)_2 \rightarrow 2MgO(s) + 4NO_2(g) + O_2(g)$

Going down Groups 1 and 2, nitrates become more stable. Higher temperatures are needed to decompose them.

Worked example

The diagram shows apparatus that can be used to investigate the thermal stability of Group 2 carbonates. The time taken for the limewater to turn cloudy is measured for each carbonate.

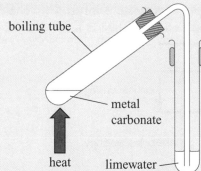

- boiling tube
- metal carbonate
- heat
- limewater

State *three* factors to control so a fair comparison can be made. **(3 marks)**

The same number of moles of carbonate should be used each time. The size of the flame and its distance to the boiling tube should be the same.

Other factors include the volume of limewater and the particle size of the carbonates. There should also be a way to standardise the measurement of cloudiness, such as a black cross drawn on the opposite side of the tube.

 Practical skills You need to understand experimental procedures to show patterns in the thermal decomposition of Group 1 and 2 nitrates and carbonates.

Explaining trends

Cations (positively charged ions) can affect anions such as CO_3^{2-} and NO_3^-.
They can lower the energy needed to break a C–O bond or N–O bond.
This effect increases:

- the smaller the cation
- the greater the cation's charge.

Li^+ ions and Group 2 ions can cause:

- CO_3^{2-} ions to decompose to O^{2-} and CO_2.
- NO_3^- ions to decompose to O^{2-}, NO_2 and O_2.

Now try this

Lithium nitrate behaves differently from other Group 1 nitrates.
(a) Describe, with the help of equations, the differences in the thermal decomposition of lithium nitrate and potassium nitrate. **(4 marks)**
(b) Explain the difference in observed thermal stability of these two nitrates. **(2 marks)**

Flame tests

Group 1 and 2 ions produce different characteristic flame colours due to electron transitions.

Flame colours for Group 1 compounds

Cation	Flame colour
Li^+	red
Na^+	yellow–orange
K^+	lilac
Rb^+	red–violet
Cs^+	blue

Sodium gives a very persistent flame colour.

Rubidium and caesium were named for their flame colours (*rubidus* means deep red and *caesius* means blue–grey in Latin).

Flame colours for Group 2 compounds

Cation	Flame colour
Be^{2+}	none
Mg^{2+}	none
Ca^{2+}	yellow–red
Sr^{2+}	red
Ba^{2+}	pale green

Beryllium and magnesium compounds do not give a colour in flame tests, but magnesium metal does burn in air with a bright white light.

Worked example

Describe how to carry out a flame test to confirm that a white powder contains potassium ions.

(3 marks)

Dip a nichrome wire loop into concentrated hydrochloric acid, then into the powder. Hold the loop in the edge of a roaring Bunsen burner flame. If the powder contains potassium ions, a lilac flame colour should be seen.

 Practical skills

Ceramic rods may be used instead, or wire loops made from platinum, nickel or chrome.

The photo shows the flame test result for potassium ions.

Electron transitions

An electron in its **ground state** is at its lowest energy level.

An electron can absorb energy when an atom or ion is heated, and becomes **excited** – it moves to a higher energy level.

An excited electron emits **electromagnetic** radiation when it falls to a lower energy level.

You see a flame colour if the frequency of the radiation is in the **visible** range.

Magnesium ions do not produce flame colours because the frequencies of the radiation emitted lie outside the visible range

Electronic configurations

When you write an electronic configuration for an atom or ion, you are showing the location of each electron in its ground state.

Different colours

An excited electron may return to its ground state through different **transitions**.

For example, it may be excited from shell 1 to 3, but de-excited from shell 3 to 2, then 2 to 1.

shell 4 ———

shell 3 ——— emits energy

shell 2 ———

shell 1 ——— absorbs energy

Each transition emits energy which corresponds to a particular colour of light.

Ionisation

An atom is **ionised** if an electron absorbs enough energy to leave it altogether.

Now try this

1 Explain, in terms of electronic transitions, how colours in a flame are produced. **(3 marks)**

2 Which of these acids is used for mixing with a salt to carry out a flame test? **(1 mark)**
   ☐ **A** concentrated nitric acid ☐ **B** concentrated sulfuric acid
   ☐ **C** concentrated hydrochloric acid ☐ **D** concentrated phosphoric acid.

Properties of Group 7 elements

The Group 7 elements include fluorine, chlorine, bromine and iodine. See the Periodic Table on page 151.

Melting and boiling temperatures

Going down Group 7:

- The melting temperatures increase.
- The boiling temperatures increase.

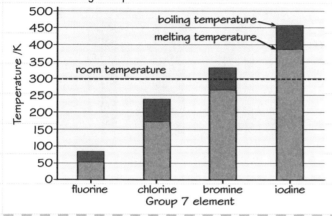

Breaking London forces

The Group 7 elements (apart from astatine):

- are reactive non-metals
- are coloured gases or form coloured vapours
- exist as diatomic molecules, e.g. Cl_2
- have London forces between their molecules.

During melting or boiling:

✓ London forces between molecules break.

✓ Covalent bonds between atoms within the molecules do not break.

Going down Group 7:

- The number of electrons in each molecule increases.
- The strength of the London forces increases.
- More energy is needed to separate molecules.

Electronegativity decreases down Group 7

Electronegativity is the ability of an atom to attract the bonding electrons in a covalent bond.

Fluorine is the most electronegative element in the Periodic Table (its electronegativity is 4.0).

Electronegativity decreases down Group 7.

This is because, although the nuclear charge increases:

- the amount of **shielding** increases
- the **distance** between the nucleus and the bonding pair in a covalent bond increases.

Appearance of the halogens

Element	Appearance at room temperature
fluorine	very pale yellow gas
chlorine	yellow–green gas
bromine	orange–brown liquid
iodine	shiny, violet-black crystals

When warmed, sublimes to form a purple vapour.

The only liquid non-metal at room temperature, but vaporises easily.

Worked example

Astatine is a rare element placed below iodine in Group 7. Use your knowledge of the other elements in Group 7 to predict the physical properties of astatine, giving reasons for your answers. **(3 marks)**

Astatine will be a solid, because melting and boiling temperatures increase down the group and iodine is a solid at room temperature. Its electronegativity will be less than that of iodine, as electronegativity decreases down the group.

You may also be asked to make predictions about the properties of fluorine at the top of Group 7.

You might also predict that astatine will have a dark colour, as the depth of colour increases down Group 7.

Its melting temperature is 575 K, its boiling temperature is 653 K, and its electronegativity is 2.2 (it is 2.5 for iodine).

Astatine is thought to be a semi-metal, with properties of a metal and a non-metal.

Now try this

1. The discovery of ununseptium, Uus, was announced in 2010. It is an artificial element placed below astatine in Group 7, but very few atoms of it have been made.
   Predict the physical properties of ununseptium, giving reasons for your answers. **(3 marks)**
2. State the type of structure and bonding present in solid iodine and explain why it easily forms a vapour when warmed. **(4 marks)**

Reactions of Group 7 elements

Group 7 elements undergo redox reactions with metals in Groups 1 and 2, and with Group 7 compounds.

Reactivity decreases down the group

Group 7 elements form negatively charged ions in reactions with Group 1 and 2 metals:

$$X_2 + 2e^- \rightarrow 2X^-$$

Group 7 elements are called **halogens** and their ions are **halide ions**.

This can be broken down into two processes:
- $X - X \rightarrow 2X$ (endothermic)
- $2X + 2e^- \rightarrow 2X^-$ (exothermic)

Going down Group 7 from chlorine to iodine:
- Energy needed to break bonds to form halogen atoms decreases.
- Energy released when halide ions form decreases.
- The overall energy change becomes less negative.
- The reactivity decreases.

Fluorine reacts vigorously with water, so its reactions in solution are difficult to observe.

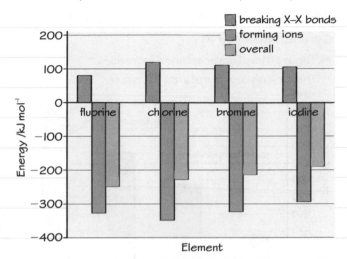

Fluorine

The F–F bond is relatively short. The three lone pairs of electrons on each fluorine atom repel strongly, making the F–F bond weaker than expected.

Oxidation reactions with Group 1 metals

The Group 7 elements react with Group 1 elements to form salts in a redox reaction.

For example, when sodium reacts with chlorine:
- Sodium is oxidised: $2Na \rightarrow 2Na^+ + 2e^-$
- Chlorine is reduced: $Cl_2 + 2e^- \rightarrow 2Cl^-$

There is a change in oxidation number:
- Sodium: $0 \rightarrow +1$ (oxidation)
- Chlorine: $0 \rightarrow -1$ (reduction)

Oxidation reactions with Group 2 metals

The Group 7 elements react with Group 2 elements to form salts in a redox reaction.

For example, when calcium reacts with chlorine:
- Calcium is oxidised: $Ca \rightarrow Ca^{2+} + 2e^-$
- Chlorine is reduced: $Cl_2 + 2e^- \rightarrow 2Cl^-$

Chlorine is acting as an **oxidising agent** when it reacts with Group 1 and 2 metals because it is gaining electrons.

Worked example

(a) Describe what is seen when chlorine is bubbled through potassium iodide solution, followed by the addition of cyclohexane.   **(2 marks)**

The solution turns brown. Two layers are then produced, and the top layer is purple.

(b) Write an ionic equation for the reaction.   **(1 mark)**

$$Cl_2 + 2I^- \rightarrow 2Cl^- + I_2$$

Halogens are more soluble in cyclohexane than in water. The displaced iodine dissolves in the upper (cyclohexane) layer, where its colour can more easily be seen. The other colours are:
- chlorine – pale green
- bromine – orange

Now try this

(a) Describe what is observed when chlorine solution is added to sodium bromide solution, followed by the addition of cyclohexane.   **(2 marks)**

(b) Write an ionic equation for the reaction observed in aqueous solution.   **(1 mark)**

(c) State and explain the role of chlorine in this reaction.   **(2 marks)**

Reactions of chlorine

In a **disproportionation reaction**, an element in a single species is simultaneously oxidised and reduced.

Reaction with water

Chlorine dissolves to form a very pale green solution often called **chlorine water**.

Some of the dissolved chlorine reacts with the water to produce a mixture of acids:

- hydrochloric acid, $HCl(aq)$
- chloric(I) acid, $HClO(aq)$.

```
        0 to +1 (oxidation)
     ┌──────────────────────┐
Cl₂(aq) + H₂O(l) ⇌ HCl(aq) + HClO(aq)
     └────────────────────↑
        0 to −1 (reduction)
```

$$Cl_2(aq) + H_2O(l) \rightleftharpoons HCl(aq) + HClO(aq)$$

It is a disproportionation reaction because chlorine is simultaneously:

- oxidised to chlorate(I) ions, ClO^-
- reduced to chloride ions, Cl^-.

Chlorine is used in water treatment

Chlorine is used to kill microorganisms in drinking water and swimming pool water.

Chloric(I) acid dissociates in water:

$$HClO(aq) \rightleftharpoons H^+(aq) + ClO^-(aq)$$

The chlorate(I) ion is a powerful **disinfectant**.

It can break down:

- bacterial cell walls
- enzymes.

Reversible reactions

The double arrow symbol in equations shows that a reaction is **reversible** – it does not go to completion and may reach equilibrium.

See page 73 for more about these reactions.

Reaction with cold alkali

Chlorine reacts with cold, dilute sodium hydroxide solution to form sodium chlorate(I):

$$Cl_2(aq) + 2NaOH(aq)$$
$$\rightarrow NaCl(aq) + NaClO(aq) + H_2O(l)$$

This is a disproportionation reaction because chlorine is simultaneously:

- oxidised to chlorate(I) ions, ClO^-
- reduced to chloride ions, Cl^-.

Sodium chlorate(I) is **bleach**.

Household bleach is used to clean and disinfect toilets, and surfaces in kitchens and bathrooms.

Reaction with hot alkali

Chlorine reacts with hot, concentrated sodium hydroxide solution to form sodium chlorate(V):

$$3Cl_2(aq) + 6NaOH(aq)$$
$$\rightarrow 5NaCl(aq) + NaClO_3(aq) + 3H_2O(l)$$

This is a disproportionation reaction because chlorine is simultaneously:

- oxidised to chlorate(V) ions, ClO_3^-
- reduced to chloride ions, Cl^-.

Worked example

Sodium chlorate(I) breaks down to form sodium chloride and sodium chlorate(V) when heated above about 75 °C. Write an ionic equation for this reaction, and explain why it is an example of disproportionation.   **(3 marks)**

$$3ClO^- \rightarrow 2Cl^- + ClO_3^-$$

Chlorine is simultaneously reduced to chloride ions (oxidation number changes from 0 to −1) and oxidised to chlorate(V) ions (oxidation number changes from 0 to +5).

The sodium ions are spectator ions and should be left out of the ionic equation.

Notice that:

☑ The total charge is the same on both sides.

☑ Total decrease in oxidation number = total increase in oxidation number

The oxidation number of oxygen stays the same, so you just need to consider chlorine here:

- decrease $2ClO^- \rightarrow 2Cl^-$ is 4: 2(+1) to 2(−1)
- increase $ClO^- \rightarrow ClO_3^-$ is 4: +1 to +5

Now try this

Iodine can react with hot, dilute sodium hydroxide solution: $3I_2(aq) + 6NaOH(aq) \rightarrow 5NaI(aq) + NaIO_3(aq) + 3H_2O(l)$

(a) State the name of the product, $NaIO_3$.   **(1 mark)**

(b) Identify the type of reaction involved, and explain your answer.   **(3 marks)**

Halides as reducing agents

Halide ions can act as **reducing agents** – species that can lose electrons.

Opposite trends

Do not confuse the oxidising ability of the halogens with the reducing ability of their ions.

Halogen	Oxidising ability	Reducing ability	Halide ion
F_2	high	low	F^-
Cl_2	↓	↑	Cl^-
Br_2			Br^-
I_2	low	high	I^-

When halide ions act as reducing agents, they lose electrons to form halogen molecules.

For example: $2Cl^- \rightarrow Cl_2 + 2e^-$

Reduction products of sulfuric acid

The oxidation number of sulfur in sulfuric acid, H_2SO_4, is +6.

There are three possible reduction products of sulfuric acid.

Name	Formula	Oxidation number of S
sulfur dioxide	SO_2	+4
sulfur	S	0
hydrogen sulfide	H_2S	−2

Reactions of solid Group 1 halides with concentrated sulfuric acid

Misty white fumes are always produced, due to an **acid–base** reaction involving hydrogen ions from the sulfuric acid.

For example:

The oxidation number of Cl^- stays at −1.

$NaCl + H_2SO_4 \rightarrow NaHSO_4 + HCl$

Cl^- ions cannot reduce sulfur in sulfuric acid.

Br^- ions can reduce sulfur in sulfuric acid to form sulfur dioxide.

I^- ions can reduce sulfur in sulfuric acid to form sulfur dioxide and also sulfur and hydrogen sulfide.

Halide	Observations	Product	Oxidation number
NaCl	misty fumes	HCl	no change
NaBr	misty fumes	HBr	no change
	brown fumes	Br_2	−1 → 0
	colourless gas with choking smell	SO_2	+6 → +4
NaI	misty fumes	HI	no change
	black solid and purple fumes	I_2	−1 → 0
	colourless gas with choking smell	SO_2	+6 → +4
	yellow solid	S	+6 → 0
	colourless gas with rotten egg smell	H_2S	+6 → −2

Worked example

(a) Write an equation to show the production of sulfur dioxide in the reaction between sodium bromide and concentrated sulfuric acid. **(2 marks)**

$2HBr + H_2SO_4 \rightarrow 2H_2O + SO_2 + Br_2$

The two half-equations are:
$2Br^- \rightarrow Br_2 + 2e^-$
$H_2SO_4 + 2H^+ + 2e^- \rightarrow 2H_2O + SO_2$

These two equations are easily combined because they both contain two electrons.

(b) Write an equation to show the production of sulfur in the reaction between sodium iodide and concentrated sulfuric acid. **(2 marks)**

$6HI + H_2SO_4 \rightarrow 4H_2O + S + 3I_2$

The two half-equations are:
$2I^- \rightarrow I_2 + 2e^-$
$H_2SO_4 + 6H^+ + 6e^- \rightarrow 4H_2O + S$

The first equation must be multiplied by 3 all the way through so that both contain six electrons.

Now try this

1 Balance this half-equation for the production of hydrogen sulfide. **(1 mark)**

$H_2SO_4 + ... H^+ + ... e^- \rightarrow ... H_2O + H_2S$

2 Write an equation to show the production of hydrogen sulfide in the reaction between sodium iodide and concentrated sulfuric acid. **(2 marks)**

Other reactions of halides

Reactions of halide ions and hydrogen halides that are not redox reactions are described here.

Testing for halide ions in solution

Silver fluoride is soluble, but the other silver halides are insoluble and form **precipitates**. This is the basis of a test for halide ions in solution:

* the sample is acidified with dilute nitric acid
* silver nitrate solution is added.

Halide	Precipitate	
	Identity	Colour
Cl^-	AgCl	white
Br^-	AgBr	cream
I^-	AgI	yellow

You could remember these colours as 'milk, cream, butter'.

The dilute nitric acid prevents other anions, such as CO_3^{2-}, forming precipitates in the tests.

Confirmatory tests

Dilute ammonia and concentrated ammonia solutions are used in **confirmatory tests**.

This is to be sure of the identity of the halide ion – you may find it difficult to tell apart the colours of faint precipitates.

Halide	Change in precipitate when ammonia solution is added	
	Dilute	Concentrated
Cl^-	dissolves	dissolves
Br^-	no change	dissolves
I^-	no change	no change

Reactions with water

The hydrogen halides:

* are colourless gases
* exist as polar molecules
* dissolve in water to form acidic solutions.

Hydrogen fluoride produces hydrofluoric acid:

$$HF(g) + H_2O(l) \rightleftharpoons H_3O^+(aq) + F^-(aq)$$

This is a weak acid, but the others are strong:

* $HCl(g) + H_2O(l) \rightarrow H_3O^+(aq) + Cl^-(aq)$
  hydrochloric acid (from hydrogen chloride)
* $HBr(g) + H_2O(l) \rightarrow H_3O^+(aq) + Br^-(aq)$
  hydrobromic acid
  (from hydrogen bromide) — oxonium ion.

Do not confuse HCl(g) with HCl(aq) (the acid).

Solubility of silver halides in ammonia

You will see a trend in the solubility of the silver halides in ammonia above. In terms of solubility:

$$AgCl > AgBr > AgI$$

A soluble ion forms when silver halides dissolve in ammonia solution. For example:

$$AgCl(s) + 2NH_3(aq) \rightarrow [Ag(NH_3)_2]^+(aq) + Cl^-(aq)$$

Reactions with ammonia

The hydrogen halides react with ammonia gas to produce ammonium salts. For example:

$$NH_3(g) + HCl(g) \rightarrow NH_4Cl(s)$$

The ammonium chloride formed is a white, ionic solid (like the other ammonium halides).

Worked example

(a) Name the acid formed when hydrogen iodide dissolves in water. **(1 mark)**

Hydriodic acid.

 Hydriodic acid is also called hydroiodic acid.

(b) Write an equation to show how it forms from hydrogen iodide. **(1 mark)**

$$HI(g) + H_2O(l) \rightarrow H_3O^+(aq) + I^-(aq)$$

 Hydrogen iodide has the state symbol (g), and an arrow is used, rather than the reversible symbol used for hydrogen fluoride.

Now try this

1 Acidified silver nitrate solution is used to test for the presence of halide ions.
   (a) Explain why fluoride ions cannot be detected using this test. **(1 mark)**
   (b) In one such test, a faint white precipitate is formed.
   (i) Explain why this may indicate the presence of chloride ions. **(1 mark)**
   (ii) Describe how you could confirm the presence of chloride ions. **(2 marks)**

2 Hexane reacts with bromine solution in the presence of ultraviolet light. Suggest, with the help of an equation, why a white smoke forms when a bottle of ammonia solution is held to the mouth of the test tube. **(3 marks)**

Exam skills 3

This exam-style question uses knowledge and skills you have already revised. Have a look at pages 28–30 for a reminder about **reactions of Group 7** and **halides as reducing agents**.

Worked example

(a) A solution called chlorine water forms when chlorine is bubbled through water.

   (i) State the colour of chlorine water. **(1 mark)**

Chlorine water is very pale green.

   (ii) Use the equation to show that a disproportionation reaction occurs:

$$Cl_2(g) + H_2O(l) \rightleftharpoons HCl(aq) + HClO(aq)$$

(2 marks)

Chlorine atoms are simultaneously reduced from oxidation number 0 to −1 in Cl^- and oxidised from 0 to +1 in ClO^-.

(b) Chlorine water is added to potassium bromide solution.

   (i) State the colour of the solution formed. **(1 mark)**

Orange.

   (ii) Write the ionic equation for the reaction, including state symbols. **(2 marks)**

$Cl_2(aq) + 2Br^-(aq) \rightarrow 2Cl^-(aq) + Br_2(aq)$

(c) Concentrated sulfuric acid is added to potassium chloride. Steamy fumes are formed, which react with ammonia to form a white smoke.

   (i) Name the gas given off. **(1 mark)**

Hydrogen chloride.

   (ii) Explain why the steamy fumes form. **(1 mark)**

Hydrogen chloride reacts with water vapour in the air.

   (iii) Name the substance in the white smoke formed when the steamy fumes react with ammonia. **(1 mark)**

Ammonium chloride, NH_4Cl.

(d) Concentrated sulfuric acid reacts with solid potassium iodide as shown in the equation:
$$8KI + 9H_2SO_4 \rightarrow 4I_2 + 8KHSO_4 + H_2S + 4H_2O$$

   State two observations made when this occurs. **(2 marks)**

There is a smell of bad eggs and a purple vapour is given off.

Command word: State

If a question asks you to **state** something, it means you need to recall one or more pieces of information.

Chlorine water is almost colourless. Remember that 'clear' is not a colour.

$HCl(aq)$ and $HClO(aq)$ are ionised in solution to form $H^+(aq)$ ions, $Cl^-(aq)$ ions and $ClO^-(aq)$ ions.

The oxidation numbers of each species are clear and it is explained that reduction and oxidation of Cl atoms occurs.

Potassium bromide solution is colourless. The bromine solution produced in the reaction may also appear to be yellow, red-brown or brown.

All the species are in aqueous solution here, even though bromine is a liquid in its standard state. $K^+(aq)$ ions are spectator ions, so are not included.

Command word: Name

If a question asks you to **name** something, it means you need to recall a piece of information.

The gas is $HCl(g)$ not $HCl(aq)$, which would be hydrochloric acid.

The hydrogen chloride gas is very soluble in water, so it attracts moisture in the air and forms hydrochloric acid droplets.

If you give a name and a formula when asked for one of these, make sure both are correct (not ammonia chloride or NH_3Cl, for example).

Iodide ions are acting as a reducing agent. They are stronger reducing agents than chloride ions or bromide ions.

You could mention that the white solid becomes a black or purple solid. Although you would see misty fumes, these are not relevant to the equation given.

Moles and molar mass

The **mole**, abbreviated to mol, is the unit for the *amount* of substance.

Defining the mole

One mole is :

• the amount of substance that

• contains the same number of particles

• as the number of carbon atoms

• in exactly 12 g of ^{12}C.

> **Be careful!** You may be used to talking about the 'amount' of a substance when you really mean its volume or mass. In Chemistry the word 'amount' has a very particular meaning, so do not muddle moles with cm³ or grams.

How many things in a mole?

The mole is defined in terms of the number of carbon atoms in 12 g of ^{12}C.

This number is the **Avogadro constant**.

The Avogadro constant, L, is:

• very large, so not used in everyday life

• approximately 6.02×10^{23} mol⁻¹

> **Maths skills** In standard form, a number is written as $a \times 10^n$
>
> • a is a number between 1 and 10
>
> • n is an integer (whole number)
>
> For example, 1.25×10^3 is 1250, and
> $L = 602\,000\,000\,000\,000\,000\,000\,000\,000$

Worked example

Assuming that the Avogadro constant is 6.0×10^{23} mol⁻¹, the number of atoms in 1 mol of water, H_2O is:

☐ **A** 2.0×10^{23} ☐ **B** 6.0×10^{23}

☐ **C** 1.2×10^{24} ☒ **D** 1.8×10^{24} **(1 mark)**

> **Be careful!** Make sure you know which particles you are applying the mole to, such as electrons, atoms, ions or molecules.

A_r and M_r

Relative atomic mass, A_r, is the weighted mean mass of an atom of an element compared to $\frac{1}{12}$th the mass of a ^{12}C atom.

Relative formula mass, M_r, is calculated by adding together all the A_r values for all the atoms present in a unit of a substance.

For example,
M_r of $H_2O = 1.0 + 1.0 + 16.0 = 18.0$

> **Be careful!** The term **relative molecular mass** is only applied to substances that exist as molecules, even though it also has the symbol M_r.

Molar mass

Molar mass:

• is the mass per mole of a substance

• has the units g mol⁻¹

The molar mass of a substance can be found from its A_r or M_r.

For example, the M_r of water is 18.0 so its molar mass is 18.0 g mol⁻¹.

Moles, mass and molar mass

This expression links these three quantities:

$m = nM$ where m = mass/ g
                     n = amount/ mol
                     M = molar mass/ g mol⁻¹

The mass of 2 mol of water = $2 \times 18.0 = 36.0$ g

Now try this

1 (a) Calculate the amount of atoms (in mol) in 2 mol of ethane, C_2H_6. **(1 mark)**

   (b) Use your answer to calculate the number of atoms in 2 mol of ethane. **(1 mark)**

2 Explain the difference between a mole and the Avogadro constant. **(1 mark)**

3 Calculate the mass (in g) of:
   (a) 0.5 mol of water, H_2O. **(1 mark)**
   (b) 1.25 mol of ethanol, C_2H_5OH. **(2 marks)**   *Calculate the M_r first.*

4 Which contains more atoms, 4.5 g of water or 3.8 g of ethanol? Explain your answer. **(2 marks)**

> *Calculate the number of moles of atoms in each substance.*

Empirical and molecular formulae

The empirical formula of a compound can be determined from experimental data.

Molecular formulae

The **molecular formula** of a substance shows the actual numbers of atoms of each element in its molecules.

For example, the molecular formula of butane is C_4H_{10}. Each molecule contains:

- four carbon atoms
- ten hydrogen atoms.

Structural formulae

You can also represent organic compounds such as butane using **structural formulae**.

These show each atom and its position.

For example, for butane it is $CH_3CH_2CH_2CH_3$.

Empirical formulae

The **empirical formula** of a compound is the simplest whole number ratio of the atoms of each element it contains.

For example, the empirical formula of butane is C_2H_5. This is because each number in its molecular formula can be simplified by dividing by 2.

Ionic compounds and giant molecules

Their formulae are usually empirical formulae:

- ✓ NaCl for sodium chloride
- ✓ SiO_2 for silicon(IV) oxide
- ✓ C for graphite.

Using molar mass

You can work out the molecular formula for a substance if you know its:

- empirical formula
- molar mass.

For example, the empirical formula for butane is C_2H_5 and its molar mass is $58\,g\,mol^{-1}$.

- molar mass for $C_2H_5 = (2 \times 12.0) + (5 \times 1.0) = 29.0\,g\,mol^{-1}$
- factor to apply = $58.0 \div 29.0 = 2$
- molecular formula is C_4H_{10} (empirical formula × 2)

Worked example

15.0 g of a lead oxide is formed when 13.6 g of lead is heated in air. Calculate its empirical formula.

(3 marks)

mass of oxygen in the oxide = 15.0 – 13.6
= 1.4 g

Symbol	Pb	O
Mass/g	13.6	1.4
A	207.2	16.0
Mass/g ÷ A	13.6 ÷ 207.2 = 0.0656	1.4 ÷ 16.0 = 0.0875
Divide by smallest value	$\frac{0.0656}{0.0656} = 1$	$\frac{0.0875}{0.0656}$ = 1.33
Adjust	3 × 1 = 3	3 × 1.33 = 4

Empirical formula is Pb_3O_4

You might be given percentage composition in a question. In that case, assume a total mass of 100 g, so 1% = 1 g.

Calculate the missing mass (or percentage) if there is one to find.

Set out the answer in columns. Use the elements' symbols as headings, and underneath them write the mass (g) and A_r values.

Calculate mass (g) ÷ A_r for each element.

Divide each number by the smallest number then adjust, if needed, to find the simplest whole number ratio. Remember the formula!

Now try this

1 The empirical formula of a compound is NO_2 and its molar mass is $92.0\,g\,mol^{-1}$. What is its molecular formula?

(1 mark)

2 A compound contains 32.4% sodium and 22.6% sulfur. The remainder is oxygen. Calculate its empirical formula.

(3 marks)

Reacting masses calculations

Mass is **conserved** in chemical reactions – the total mass of reactants and products stays the same.

Moles, mass and molar mass

This formula links these three quantities:

$n = \dfrac{m}{M}$, where n = amount/ mol

m = mass/ g
M = molar mass/ g mol^{-1}

The amount of water in 36.0 g (molar mass 18.0 g mol^{-1}) = $36.0 \div 18.0 = 2.0$ mol

Remember that you can calculate the molar mass of a substance using:

- A_r values for each element it contains (given in the Periodic Table)
- the formula of the substance.

Equations

An equation is **balanced** when it has the same number of atoms of each element on the left and right sides of the equation.

For example:

- $N_2 + H_2 \rightarrow NH_3$ is not balanced
- $N_2 + 3H_2 \rightarrow 2NH_3$ is balanced

The unbalanced equation implies that, in the reaction, nitrogen atoms disappear and hydrogen atoms appear.

This cannot happen, because it breaks the principle of **conservation of mass**.

Balanced equations obey this principle.

Worked example

Calculate the mass of ammonia that can be formed from 12.0 g of hydrogen:

$N_2 + \underline{3H_2} \rightarrow \underline{2NH_3}$ **(4 marks)**

molar masses:

$H_2 = 2.0$ g mol^{-1} and $NH_3 = 17.0$ g mol^{-1}

amount of hydrogen = $12.0 \div 2.0 = 6.0$ mol

amount of ammonia = $2 \div 3 \times 6.0 = 4.0$ mol

mass of ammonia = $4.0 \times 17.0 = 68.0$ g

It may help if you underline the two substances given in the question so you focus on them.

Follow these steps:

☑ Calculate the molar masses needed.

☑ You will have the mass and molar mass of one substance – use these to calculate its amount.

☑ Work out the ratio of the two substances in the equation ($2 \div 3$ here) and then the amount of the second substance.

☑ Calculate the mass of the second substance.

Worked example

When an oxide of copper is heated in hydrogen to constant mass, 3.2 g of copper and 0.90 g of water are formed. Deduce the equation for the reaction. **(4 marks)**

molar masses:
$Cu = 63.5$ g mol^{-1} and $H_2O = 18.0$ g mol^{-1}

amount of copper = $3.2 \div 63.5 = 0.05$ mol

amount of water = $0.90 \div 18.0 = 0.05$ mol

ratio is $0.05 : 0.05$ or $1 : 1$

The equation must be:

$CuO + H_2 \rightarrow Cu + H_2O$

Copper oxide exists as:
- copper(I) oxide, Cu_2O
- copper(II) oxide, CuO

☑ Calculate the molar masses needed.
☑ Calculate the amount of each substance.
☑ Work out the simplest whole number ratio of these substances.
☑ Write the balanced equation in which the substances have this ratio.

The equation to form the other oxide is:
$Cu_2O + H_2 \rightarrow 2Cu + H_2O$

Now try this

1 What mass of magnesium chloride, $MgCl_2$, can be formed from the reaction of 5.98 g of magnesium with an excess of hydrochloric acid, HCl? Give your final answer to 3 significant figures. **(4 marks)**

2 When an oxide of copper is heated in hydrogen to constant mass, 3.51 g of copper and 0.499 g of water are formed. Deduce the equation for the reaction. **(3 marks)**

Gas volume calculations

Avogadro's law states that equal volumes of gases under the same conditions of temperature and pressure contain the same numbers of molecules.

Reacting volumes

Under given conditions of temperature and pressure, gases react together in whole number ratios of volume. For example:

$$H_2(g) \quad + \quad Cl_2(g) \quad \rightarrow \quad 2HCl(g)$$
$$100\,cm^3 \qquad 100\,cm^3 \qquad 200\,cm^3$$

The gas volumes are in the same ratio as the formulae in the equation, $1:1:2$.

This means that $50\,cm^3$ of H_2 would react with $50\,cm^3$ of Cl_2 to produce $100\,cm^3$ of HCl.

It also means that if $50\,cm^3$ of H_2 was mixed with $75\,cm^3$ of Cl_2:

- $50\,cm^3$ of the Cl_2 would react
- $25\,cm^3$ of the Cl_2 would not react (so Cl_2 would be **in excess**)
- $100\,cm^3$ of HCl would be produced
- total gas volume after the reaction would be $(100\,cm^3\ HCl + 25\,cm^3\ Cl_2) = 125\,cm^3$.

Molar volume

The volume of a fixed amount of gas increases as:

✓ the temperature increases

✓ the pressure decreases.

This means you need to state the temperature and pressure of a gas to make comparisons.

The **molar volume**, V_m, of any gas is:

- $24.0\,dm^3\,mol^{-1}$ at RTP ──room temperature and pressure (293 K and 101.325 kPa)
- $22.4\,dm^3\,mol^{-1}$ at STP.

standard temperature and pressure (273 K and 100 kPa)

Be careful!

If a reaction involves H_2O, check if it is:

✓ $H_2O(g)$ where you can apply V_m

✗ $H_2O(l)$ where you cannot apply V_m

The volume of liquid water is about 1300 times less than the volume of gaseous water at RTP.

Worked example

The complete combustion of $50\,cm^3$ of a hydrocarbon vapour gave $200\,cm^3$ of carbon dioxide (both gas volumes were measured at RTP). The formula of the hydrocarbon could be:

☐ **A** CH_4

☒ **B** C_4H_{10}

☐ **C** C_5H_{12}

☐ **D** C_8H_{18} **(1 mark)**

Hydrocarbons are compounds of hydrogen and carbon only. A general equation for their complete combustion is:

$$C_xH_y + \left(\frac{x+y}{4}\right)O_2 \rightarrow xCO_2 + \frac{y}{2}H_2O$$

So the volume ratio for hydrocarbon: CO_2 gives the value of x in the hydrocarbon.

$(200\,cm^3 \div 50\,cm^3) = 4$

So $x = 4$ and the hydrocarbon could be C_4H_{10}.

It could also be C_4H_8 but that is not an option.

Worked example

Calculate the volume, in cm^3, occupied by $0.0200\,mol$ of oxygen at 298 K and 100 kPa pressure.
            (3 marks)

$pV = nRT$

$\quad = 4.95 \times 10^{-4}\,m^3$

Volume $= 4.95 \times 10^{-4} \times 10^6 = 495\,cm^3$

This question involves the **ideal gas equation**.

This expression applies to gases and to volatile liquids above their boiling temperature:

- p = pressure in Pa
- V = volume in m^3 (care – not dm^3 or cm^3)
- n = amount of gas in mol
- R = gas constant ($8.31\,J\,mol^{-1}\,K^{-1}$)
- T = temperature in K

Multiply by 10^6 to convert m^3 to cm^3.

Now try this

Calculate the amount of carbon first, using A_r C $= 12.0\,g\,mol^{-1}$.

Carbon reacts with oxygen: $C(s) + O_2(g) \rightarrow CO_2(g)$

(a) Calculate the volume of oxygen (in dm^3 at STP) that would react with $0.536\,g$ of carbon.   **(2 marks)**

(b) Calculate the total volume of gas after the reaction, if $1.40\,dm^3$ of oxygen was present at the start.   **(1 mark)**

Concentrations of solutions

The **concentration** of solute in a solution can be measured in mol dm^{-3} or in g dm^{-3}.

Molar concentration

Molar concentration can be calculated using:

$$\text{concentration} = \frac{\text{amount of solute} \overset{\text{mol}}{}}{\text{volume of solution}}$$

mol dm^{-3} dm^3

For example, 0.3 mol of NaOH are contained in $0.5\,\text{dm}^3$ of solution.

Concentration, $c = \dfrac{n}{V} = 0.3 \div 0.5$

$\phantom{\text{Concentration, } c} = 0.6\,\text{mol dm}^{-3}$

This can be shown as $[\text{NaOH(aq)}] = 0.6\,\text{mol dm}^{-3}$.

square brackets [ ] mean
concentration in mol dm^{-3}

Mass concentration

Mass concentration can be calculated using:

$$\text{concentration} = \frac{\text{mass of solute} \overset{\text{g}}{}}{\text{volume of solution}}$$

g dm^{-3} dm^3

For example, 12 g of NaOH are contained in $0.5\,\text{dm}^3$ of solution.

$\text{Mass concentration} = \dfrac{m}{V} = 12 \div 0.5 = 24\,\text{g dm}^{-3}$

Do not use square brackets for mass concentration.

Volume measurements

In the lab, you will normally make volume measurements in cm^3 but molar concentration is measured in mol dm^{-3}.

This means that you must be able to convert from cm^3 to dm^3 (and sometimes back again):

- $\text{cm}^3 \rightarrow \text{dm}^3 \dots \times 10^{-3}$ or $\div 1000$
- $\text{dm}^3 \rightarrow \text{cm}^3 \dots \times 10^{3}$ or $\times 1000$

For example:

$25\,\text{cm}^3 = 25 \times 10^{-3}\,\text{dm}^3$ (or $0.025\,\text{dm}^3$)

Maths skills Do not mix cm^3 and dm^3 in your calculations because you will get ridiculous answers that are far too high or too low. You may sometimes get the correct answer but with incorrect working, especially in titration calculations.

Practical skills — Making volume measurements

Measuring cylinders are often used in the lab but they have high **measurement errors**.

For example, a $50\,\text{cm}^3$ measuring cylinder has graduations of $1\,\text{cm}^3$.

Its **percentage error** is $100 \times 1 \div 50 = 2\%$

A $250\,\text{cm}^3$ **volumetric flask** has a low measurement error, typically $\pm 0.2\,\text{cm}^3$ or 0.08%.

It is used to make **standard solutions**, whose concentrations are accurately known.

The **meniscus** in the solution lies on the mark in the volumetric flask.

Worked example

Calculate the mass of sodium hydroxide needed to make $500\,\text{cm}^3$ of a $0.3\,\text{mol dm}^{-3}$ solution. (Molar mass of NaOH $= 40.0\,\text{g mol}^{-1}$)   **(2 marks)**

volume $= 500 \div 1000 = 0.5\,\text{dm}^3$

amount of NaOH $= 0.3 \times 0.5 = 0.15\,\text{mol}$

mass $= 0.15 \times 40.0 = 6.0\,\text{g}$

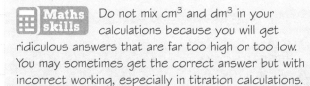

You could calculate the molar mass using the A_r values for Na, O and H given in the Periodic Table.

Remember to convert cm^3 to dm^3.

A quick way to convert from mol dm^{-3} to g dm^{-3}:
- $\text{g dm}^{-3} = \text{mol dm}^{-3} \times$ molar mass

Look at the two examples at the top of the page.

Now try this

1   0.1 mol of sodium chloride, NaCl, is present in $50\,\text{cm}^3$ of sodium chloride solution.
   (a)   Calculate the molar concentration of solution.   **(2 marks)**
   (b)   Calculate the amount of ions present in the solution.   **(1 mark)**

Sodium chloride solution contains $\text{Na}^+(\text{aq})$ ions and $\text{Cl}^-(\text{aq})$ ions.

2   What mass of glucose is in $250\,\text{cm}^3$ of a $0.1\,\text{mol dm}^{-3}$ solution? (Molar mass $= 180.0\,\text{g mol}^{-1}$).
   ☐ **A** $4.5\,\text{kg}$ ☐ **B** 4   ☐ **C** $4.5\,\text{g}$ ☐ **D** $0.45\,\text{g}$ **(1 mark)**

Doing a titration

Practical skills A **titration** is a method in which the volumes of two reacting solutions are measured, so that the concentration of one of the solutions can be determined.
You will need to demonstrate you can carry out a titration, using a burette and a pipette.

Outline of method

In an acid–alkali titration, you often put the acid into the burette and the alkali into the conical flask, but not always. The main steps are:

- Use a pipette to transfer 25 cm³ of alkali to the conical flask and add some indicator.
- Add acid from the burette to the flask until the end point is reached.
- Record the burette reading before and after adding the acid.

> The end point and equivalence point should coincide, but this does not always happen.

Key words

Word(s)	Meaning
Meniscus	the curve in the upper surface of a liquid
End point	where the indicator first changes colour
Titre	volume of solution added to reach the end point
Concordant results	when two or more titres are within 0.10 cm³ of each other
Equivalence point	where the reaction is exactly completed

Tips for accurate results

- Add only a few drops of indicator.
- Use a white tile to see the colour clearly.
- Make sure the tip of the burette is full.
- Keep the burette vertical.
- Take readings from the bottom of the meniscus.
- Record readings to ±0.05 cm³.

- Add drop-wise near the end point.
- Rinse the inside of the flask with deionised water near the end point.
- Stop adding when the indicator first changes colour.
- Repeat until you get concordant results.

Worked example

(a) Describe two precautions you would take to carry out an acid–alkali titration safely. **(2 marks)**

Use a pipette filler to fill the pipette, and do not fill the burette above eye level.

(b) Describe the colour change for phenolphthalein at the end point when acid is added to alkali in a titration. **(1 mark)**

The colour changes from red to colourless.

> The precautions for safe working are not the same as those for accurate working.

> You should not reach too high when you fill a burette. You should remove the funnel after filling, and you do not have to start at 0.00 cm³.

> Phenolphthalein is suitable for strong acid into strong base, and for weak acid into strong base.
> Methyl orange is a better choice for strong acid into weak base. It is yellow in alkali and red in acid.

Now try this

1 Describe the colour change for methyl orange at the end point when hydrochloric acid is added from a burette to ammonia solution in a conical flask. **(1 mark)**

> Hydrochloric acid is a strong acid and ammonia is a weak base.

2 The table shows four titres from a titration. Tick the concordant results. **(1 mark)**

Titre/cm³	25.30	24.85	25.15	24.95
Tick if concordant				

Titration calculations

If you know the volume and concentration of one of the solutions used in a titration, you can calculate the concentration of the other solution if you know its volume.

Outline of calculations

These are the steps you need to take:

1 Calculate the mean titre of the **titrant**. —— The standard solution, usually added from the burette.

2 Calculate the amount of solute in the titrant. —— Use its concentration and the mean titre.

3 Calculate the amount of solute in the **analyte**. —— Use your answer to step 2 and the reactant ratio from the balanced equation.
—— The unknown solution, usually put into the flask.

4 Calculate the concentration of the analyte. —— Use your answer to step 3 and the volume of analyte.

The mean titre

You need to select at least two **concordant** results for your **mean titre**. For example:

The first run is usually a trial run, carried out quickly to get an idea of the titre, and should be ignored.

Titration number	1	2	3	4
Final reading/cm^3	24.15	46.75	22.80	45.45
Initial reading/cm^3	1.20	24.15	0.00	22.80
Titre/cm^3	22.95	22.60	22.80	22.65
Concordant titres		✓		✓

You should make sure that you:

- record your readings to 2 decimal places, ending in 0 or 5
- tick the concordant titres (those within 0.10 cm^3 of each other)
- only use the concordant titres to calculate the mean titre.

In this example:

expressed to 2 decimal places, where the last number can be any number

mean titre $= \dfrac{22.60 + 22.65}{2} = 22.63\,cm^3$

The error in each burette reading is 0.05 cm^3. As a titre uses two readings, its error is 0.10 cm^3.

In the example, the percentage error in the mean titre $= \dfrac{0.10}{22.63} \times 100 = 0.44\%$

Worked example

22.40 cm^3 of 0.100 mol dm^{-3} sulfuric acid reacted with 25.00 cm^3 of a sodium hydroxide solution:

$H_2SO_4 + 2NaOH \rightarrow Na_2SO_4 + 2H_2O$

Calculate the concentration of the sodium hydroxide solution, giving your answer to 3 significant figures.
(2 marks)

amount = concentration × volume

From the equation, 1 mol of H_2SO_4 reacts with 2 mol of NaOH.

concentration = amount ÷ volume

Amount of H_2SO_4 = 0.100 × 22.40 × 10^{-3}
= 2.24 × 10^{-3} mol

Amount of NaOH = 2 × 2.24 × 10^{-3}
= 4.48 × 10^{-3} mol

Concentration of NaOH = (4.48×10^{-3}) ÷ (25.00 × 10^{-3})
= 0.179 mol dm^{-3}

Now try this

In a titration, 23.45 cm^3 of 0.080 mol dm^{-3} sulfamic acid reacted with 25.00 cm^3 of a sodium hydroxide solution: $NH_2SO_3H + NaOH \rightarrow NH_2SO_3Na + H_2O$
Calculate the concentration of sodium hydroxide.
(3 marks)

Atom economy

Percentage yield and atom economy are two ways to assess chemical reactions, particularly those used in industrial processes.

Percentage yield

This expression is used to calculate the **percentage yield** of a reaction:

$$\text{percentage yield} = \frac{\text{actual yield}}{\text{theoretical yield}} \times 100$$

The percentage yield is usually less than 100% because:

- the reaction may be reversible and not go to completion
- side reactions may lead to by-products
- product may be lost during purification.

Some definitions

☑ **Desired product** – the useful substance you want to make.

☑ **Yield** – the mass of desired product obtained in a reaction.

☑ **Actual yield** – the measured yield from a particular reaction.

☑ **Theoretical yield** – the maximum possible yield determined from reacting mass calculations for a reaction.

Atom economy

The **atom economy** of a reaction is a measure of how many atoms of the reactants end up in the desired product:

— its total molar mass

$$\text{atom economy} = \frac{\text{molar mass of the desired product}}{\text{sum of the molar masses of all products}} \times 100$$

— or of all reactants

In industrial processes:

- a high atom economy means there is less waste to dispose of
- the atom economy can be improved if a by-product can be sold, rather than just disposed of.

In general:

- Addition reactions have 100% atom economy.

— A reaction with 100% atom economy may not have 100% percentage yield.

- Elimination, substitution and multistep reactions have lower atom economies.

Worked example

Ethanol, C_2H_5OH, can be manufactured by the fermentation of glucose:

$$C_6H_{12}O_6 \rightarrow 2C_2H_5OH + 2CO_2$$

(a) Calculate the atom economy. **(2 marks)**

molar mass of $C_2H_5OH = 46.0\,g\,mol^{-1}$
molar mass of $CO_2 = 44.0\,g\,mol^{-1}$

$$\text{sum of molar masses} = (2 \times 46.0) + (2 \times 44.0)$$
$$= 180.0$$

$$\text{atom economy} = \frac{(2 \times 46.0)}{180.0} \times 100 = 51.1\%$$

The total molar mass of the desired product is 2×46.0 because it is $2C_2H_5OH$ in the equation.

(b) 2.30 g of ethanol was obtained from 36.0 g of glucose. Calculate the percentage yield. **(2 marks)**

molar mass of $C_6H_{12}O_6 = 180.0\,g\,mol^{-1}$

$$\text{theoretical yield} = \frac{36.0}{180.0} \times 2 \times 46.0$$
$$= 18.4\,g$$

$$\text{percentage yield} = \frac{2.30}{18.4} \times 100 = 12.5\%$$

The atom economy could be improved to 100% if the CO_2 was also desirable. For example, it could be sold to make drinks fizzy.

Now try this

Silver chloride can be made from silver nitrate solution and sodium chloride solution:

$$AgNO_3(aq) + NaCl(aq) \rightarrow NaNO_3(aq) + AgCl(s)$$

(a) Calculate the atom economy. **(2 marks)**

(b) 12.0 g of silver chloride was made from 20.0 g of silver nitrate. Calculate the percentage yield. **(2 marks)**

Exam skills 4

This exam-style question uses knowledge and skills you have already revised. Have a look at pages 37–39 for a reminder about **concentrations of solutions** and **titration calculations**.

Worked example

Magnesium hydroxide is used as an antacid, a medicine used to neutralise excess stomach acid. A crushed 1.08 g sample of an antacid was dissolved in 50.0 cm³ of 1.00 mol dm⁻³ hydrochloric acid. The mixture was added to a volumetric flask and made up to 250 cm³ with deionised water. The mean titre was 18.00 cm³ when 25.0 cm³ portions of this mixture were titrated with 0.100 mol dm⁻³ sodium hydroxide solution.

The equations for the reactions involved are:

$Mg(OH)_2(aq) + 2HCl(aq) \rightarrow MgCl_2(aq) + 2H_2O(l)$

$HCl(aq) + NaOH(aq) \rightarrow NaCl(aq) + H_2O(l)$

(a) Calculate the number of moles of hydrochloric acid that reacted with the sodium hydroxide solution. **(1 mark)**

Amount of NaOH = $0.100 \times \dfrac{18.00}{1000}$

$= 1.8 \times 10^{-3}$ mol

Amount of HCl = 1.80×10^{-3} mol

(b) Calculate the number of moles of hydrochloric acid that were left after the reaction with the tablet. **(1 mark)**

Amount of HCl = $10 \times 1.80 \times 10^{-3}$

$= 1.80 \times 10^{-2}$ mol

(c) Calculate the number of moles of hydrochloric acid added to the crushed antacid tablet. **(1 mark)**

Amount of HCl = $1.00 \times \dfrac{50.0}{1000}$

$= 5.00 \times 10^{-2}$ mol

(d) Use your answers to (b) and (c) to calculate the number of moles of magnesium hydroxide that reacted with the hydrochloric acid. **(2 marks)**

Excess amount of HCl $= (5.00 \times 10^{-2}) - (1.80 \times 10^{-2})$

$= 3.20 \times 10^{-2}$ mol

Moles of $Mg(OH)_2$ reacted $= \dfrac{(3.20 \times 10^{-2})}{2}$

$= 1.60 \times 10^{-2}$ mol

(e) Calculate the percentage of magnesium hydroxide in the tablet. Give your answer to three significant figures. **(2 marks)**

Molar mass of $Mg(OH)_2$ = 58.3 g mol⁻¹

Mass of $Mg(OH)_2$ = $1.60 \times 10^{-2} \times 58.3$

$= 0.9328$ g

Percentage $= \dfrac{0.9328}{1.08} \times 100 = 86.4\%$

Practical skills This method describes a 'back titration'.

In a sense this is a titration done in reverse. Instead of titrating the original sample, you add a known excess of a standard solution to the sample and then titrate the remaining excess.

Back titrations are useful when the sample is insoluble in water (as here), or when the end point would be difficult to determine.

Command word: Calculate

If a question asks you to **calculate** something, it means you need to:

- obtain a numerical answer
- show relevant working
- include the unit if the answer has one.

 1:1 mole ratio between NaOH and HCl.

 Remember that the mixture of acid and crushed tablet was made up to 250 cm³ with water, then 25 cm³ portions were titrated. So the total amount is $\left(\frac{250}{25}\right)$ = 10 times the amount in one titration.

 Remember to keep looking back at the stem of the question (the start of it) to see what information you can use in your answers.

$\text{amount reacted} = \left(\begin{smallmatrix}\text{amount} \\ \text{at start}\end{smallmatrix}\right) - \left(\begin{smallmatrix}\text{amount} \\ \text{at end}\end{smallmatrix}\right)$

 There is a 1:2 mole ratio between $Mg(OH)_2$ and HCl, so the amount of magnesium hydroxide is half the amount of hydrochloric acid.

Maths skills If you forget to divide by 2 in part (d), you will get an answer in part (e) that is obviously wrong (173%). This should prompt you to go back and check your working out.

 This type of calculation can also be used to determine percentage purity.

Alkanes

Hydrocarbons are compounds of hydrogen and carbon only. The alkanes are hydrocarbons.

Features of alkanes

The **alkanes** have these features in common:

- They are **saturated** hydrocarbons (they only contain single bonds).

- They have the **general formula** C_nH_{2n+2}

For example, eicosane contain 20 carbon atoms, so its **molecular formula** is: $C_{20}H_{42}$

$n = 20$, so 20 carbon atoms

$n = 20$, so $(2 \times 20) + 2 = 42$ hydrogen atoms

The molecular formula shows the number of atoms of each element in a molecule.

Names and molecular formulae

The name of an alkane ends in **-ane**.

The first part of the name refers to the number of carbon atoms its molecules contain.

C atoms	First part of name	Alkane	
		Formula	Name
1	meth	CH_4	methane
2	eth	C_2H_6	ethane
3	prop	C_3H_8	propane
4	but	C_4H_{10}	butane
5	pent	C_5H_{12}	pentane
6	hex	C_6H_{14}	hexane

Displayed and structural formulae

Shows each atom and the bonds it has with other atoms.

Shows each carbon atom and the atoms bonded to it.

Name	Displayed formula	Structural formula
methane	H │ H—C—H │ H	CH_4
ethane	H H │ │ H—C—C—H │ │ H H	CH_3CH_3
propane	H H H │ │ │ H—C—C—C—H │ │ │ H H H	$CH_3CH_2CH_3$
butane	H H H H │ │ │ │ H—C—C—C—C—H │ │ │ │ H H H H	$CH_3CH_2CH_2CH_3$
pentane	H H H H H │ │ │ │ │ H—C—C—C—C—C—H │ │ │ │ │ H H H H H	$CH_3CH_2CH_2CH_2CH_3$

The molecular and structural formulae are the same for methane.

The C–C bonds are sometimes shown, e.g. CH_3–CH_2–CH_3.

Sometimes simplified to $CH_3(CH_2)_3CH_3$ to save space.

Worked example

(a) State the empirical formula for hexane, C_6H_{14}. **(1 mark)**

The empirical formula is C_3H_7.

(b) Draw its skeletal formula. **(1 mark)**

The **skeletal formula** shows the bonds between carbon atoms in the alkane molecule.

Now try this

Octane is an alkane whose molecules each contain 8 carbon atoms.
(a) Write the molecular, empirical and structural formulae for octane. **(3 marks)**
(b) Draw the displayed and skeletal formulae for octane. **(2 marks)**

Isomers of alkanes

Alkanes with four or more carbon atoms can have **structural isomers**.

Isomers

Isomers are compounds with the same molecular formula, but different arrangements of atoms in space.

Structural isomers have the same molecular formula but their atoms are arranged in a different order:

- Chain isomers are structural isomers in which the carbon chains are different.

Chain isomers of alkanes can be:
- straight chains (unbranched)
- chains with one or more branches

The branches or side chains can be in different positions on the main chain.

Prefixes for branches

Branches are named using a **prefix**.

Each prefix is based on the number of carbon atoms in the branch.

C atoms	Prefix	Formula
1	methyl	$-CH_3$
2	ethyl	$-C_2H_5$
3	propyl	$-C_3H_7$
4	butyl	$-C_4H_9$
5	pentyl	$-C_5H_{11}$
6	hexyl	$-C_6H_{13}$

Finding structural isomers

Methane, ethane and propane do not have structural isomers, but the other alkanes do. For example, C_5H_{12} has three isomers. Start with the straight chain isomer:

H   H   H   H   H
| | | | |
H—C— C— C— C— C—H
| | | | |
H   H   H   H   H

Then imagine removing one carbon atom and attaching it somewhere else on the chain. It helps at this stage just to show the carbon chain.

C— C—C— C
        |
        C

These chains are all the same as the one above:

C
|
C— C—C— C
flipped vertically

C— C—C— C
        |
        C
flipped horizontally

C
|
C— C—C
    |
    C
main chain bent

If there are no more isomers, remove two carbon atoms and repeat.

        C
        |
C— C—C
        |
        C

No more isomers are possible here. Remember to add in the hydrogen atoms afterwards.

Naming structural isomers

For example, what is the name of this alkane?

H_3C — CH — CH — CH_2 — CH_3
       | |
      CH_3 CH_3

You first find the longest carbon chain and name it. In this case, pentane (for 5 carbon atoms):

H_3C — CH — CH — CH_2 — CH_3
       | |

You next find the branches. In this case, there are two methyl branches, so you write dimethyl:

| |
CH_3 CH_3

You then number each branch, making sure that the total of the numbers is as low as possible:

5   4   3   2   1
1   2   3   4   5
H_3C — CH — CH — CH_2 — CH_3
       | |
      CH_3 CH_3

If you number from:
- the left, the total is (2 + 3) = 5
- the right, the total is (3 + 4) = 7

So it is not 3,4-dimethylpentane.

It is 2,3-dimethylpentane.

Separate two numbers using a comma, and a number and a word using a hyphen

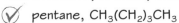

What are the names of these isomers?

These isomers, in order, are:

- ✓ pentane, $CH_3(CH_2)_3CH_3$
- ✓ 2-methylbutane, $(CH_3)_2CHCH_2CH_3$
- ✓ 2,2-dimethylpropane, $C(CH_3)_4$

Now try this

There are nine chain isomers of heptane, C_7H_{16}.
Draw the structure of each isomer, and name it.
(18 marks)

Alkenes

The alkenes are hydrocarbons whose molecules contain at least one C=C bond.

Features of alkenes

The **alkenes** have these features in common:

- They are **unsaturated** hydrocarbons (they contain at least one C=C bond).
- They have the **general formula** C_nH_{2n}.

For example, eicosene contain 20 carbon atoms, so its **molecular formula** is: $C_{20}H_{40}$

$n = 20$, so 20 carbon atoms

$n = 20$, so $(2 \times 20) = 40$ hydrogen atoms

The **empirical formula** of all alkenes with one C=C bond is CH_2.

Names and molecular formulae

The name of an alkene ends in **-ene**.

The first part of the name refers to the number of carbon atoms its molecules contain.

C atoms	First part of name	Alkane	
		Formula	Name
2	eth	C_2H_4	ethene
3	prop	C_3H_6	propene
4	but	C_4H_8	butene
5	pent	C_5H_{10}	pentene
6	hex	C_6H_{12}	hexene

Representing alkenes

	Name	Displayed formula	Structural formula	Skeletal formula
The bond angles are close to 120° (the H–C–H bond angle is actually 117.3°).	ethene		$CH_2=CH_2$	
It is often easier to show the carbon–carbon bonds in a straight line.	propene		$CH_3CH=CH_2$	

The skeletal formula for propene is not in a straight line.

Worked example

Two hydrocarbons (A and B) have the molecular formula C_5H_{10}. Hydrocarbon A decolourises bromine water but hydrocarbon B does not. Explain this observation, and suggest a structure and name for hydrocarbon B.   **(3 marks)**

Hydrocarbon A must be unsaturated but hydrocarbon B must be saturated. It could be cyclopentane.

Bromine water (an aqueous solution of bromine) is used in a **qualitative test** for the presence of C=C bonds. In the test:

- Bromine water is added to the hydrocarbon.
- It turns from orange-brown to colourless if the hydrocarbon is unsaturated.

Bromine may be used but it is more hazardous.

Cyclopentane is a **functional group isomer** of pentene, C_5H_{10}. It has the same molecular formula but a different functional group.

Functional groups determine the main chemical properties of an organic compound:
- C=C is the functional group in alkenes.

Now try this

1 Octene is an alkene whose molecules each contain 8 carbon atoms. Write the molecular and empirical formulae for octene.   **(2 marks)**

2 Buta-1,3-diene is used in the manufacture of synthetic rubber. Suggest its displayed and skeletal formulae.   **(2 marks)**

Buta-1,3-diene contains two C=C bonds, not just one.

Isomers of alkenes

Alkenes may have **stereoisomers** because the C=C bond has restricted rotation, so the attached groups cannot change position relative to the bond.

Position isomers

Position isomers are structural isomers with the same carbon chain, but with the functional group in different places.

For example, butene has two unbranched position isomers.

but-1-ene   but-2-ene

You indicate the position of the C=C using a number (the **locant**). The locant for a C=C bond is the lowest possible, so above it is:

- but-1-ene *not* but-2-ene
- but-2-ene *not* but-3-ene

Notice how the name is split, with the locant written before ene, and separated by hyphens.

Chain isomers

Chain isomers are structural isomers in which the carbon chains are different.

For example, butene has a chain isomer.

It is called 2-methylprop-1-ene:

The longest carbon chain containing the functional group, C=C, has 3 carbon atoms in it.

This is named with the lowest locant:
- prop-1-ene

This means that the carbon atom with the methyl group attached is number 2:
- 2-methyl

Putting both parts together gives you:
- 2-methylprop-1-ene

It is not, for example, 2-methylprop-2-ene.

Stereoisomers

Stereoisomers have:
- ✓ the same molecular formula
- ✓ the same structural formula
- ✓ a different arrangement of atoms.

E–Z isomers

But-2-ene has two stereoisomers:

E-but-2-ene   Z-but-2-ene

Each of the C=C carbon atoms has a –CH$_3$ group and an H atom attached:
- –CH$_3$ has a **higher priority** because its total atomic number is higher (15 rather than 1).

In the *E* isomer, the higher priority groups are opposite each other (German: *entgegen*).

In the *Z* isomer, the higher priority groups are together (German: *zusammen*).

Be careful!

One way to remember *E–Z* isomers is:

E = Either side

Z = Zame zide

cis–trans isomers

Substituent groups are atoms or groups attached to the carbon atoms in the C=C bond.

If the substituent groups on both carbon atoms are the same, the stereoisomers can also be named using the *cis–trans* notation:
- *E*-but-2-ene is also called *trans*-but-2-ene
- *Z*-but-2-ene is also called *cis*-but-2-ene

The functional groups look like they form a C on its side.

For an alkene to have stereoisomers:
- Each carbon atom involved in the C=C bond must have two different groups attached.

So but-1-ene does not have stereoisomers:
- one of the C=C carbon atoms has two H atoms attached to it.

Now try this

1 Name this alkene.

(1 mark)

2 Pent-2-ene has two stereoisomers. Name each one, and draw its displayed formula.   **(4 marks)**

Using crude oil

Crude oil (petroleum) is a mixture of hydrocarbons that can be separated by fractional distillation. Each fraction produced contains hydrocarbons with similar boiling points and chain lengths.

Fractional distillation

Fractional distillation relies on differences in the boiling temperatures of the different hydrocarbons in crude oil. In general:

- The boiling temperature increases as the number of carbon atoms in the alkane molecules increases.

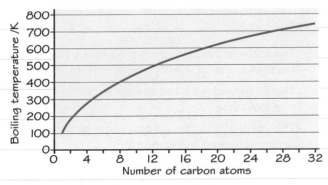

Fractions

The crude oil is heated and the vapours are led into a fractionating column, which is hot at the bottom and cool at the top.

As the vapours rise, they cool and condense:
- Liquids are drawn off at various heights.
- Refinery gas does not condense, but leaves from the top of the column.

Fraction name	Boiling temperature	Size of molecules
Refinery gas	lowest	smallest
Petrol	↑	↑
Kerosene		
Diesel oil		
Fuel oil		
Bitumen	highest	largest

Leaves from the bottom of the fractionating column.

Cracking

Cracking involves chemical reactions that:

- happen at high temperatures
- use catalysts such as **zeolites**
- decompose larger alkanes to produce smaller alkanes and alkenes.

For example, for decane:

$$C_{10}H_{22} \rightarrow C_8H_{18} + C_2H_4$$

octane, used in petrol   ethene, used to make polymers

Cracking is carried out to balance supply with demand of hydrocarbons. Fractionating crude oil produces:
- too much of the larger alkanes
- not enough of the smaller alkanes.

🧪 Practical skills   Cracking in the lab

This sort of apparatus is used to demonstrate cracking. The aluminium oxide acts as a catalyst.

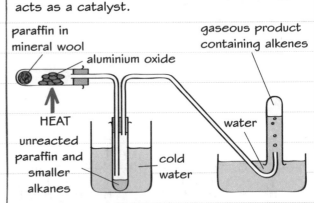

paraffin in mineral wool   aluminium oxide   gaseous product containing alkenes

HEAT   water

unreacted paraffin and smaller alkanes   cold water

To avoid suck back, the delivery tube is lifted out of the water before removing the heat.

Worked example

(a) Name the process represented by the following equation. **(1 mark)**

$+ H_2$

This is an example of reforming.

(b) Explain why this process is used. **(2 marks)**

Reforming converts unbranched hydrocarbons, which have lower octane ratings, into branched hydrocarbons and cyclic hydrocarbons. These have higher octane ratings, so they burn more efficiently in vehicle engines.

Now try this

Briefly describe each process in oil refining (fractional distillation, cracking and reforming), and outline why it is carried out. **(6 marks)**

Hydrocarbons as fuels

Combustion of alkanes is an example of an **oxidation** reaction, in which oxygen reacts with the hydrogen and carbon atoms in the alkane molecules.

Complete combustion

Complete combustion happens when there is sufficient oxygen (usually from the air) to oxidise the alkanes fully:
- Hydrogen is oxidised to water, H_2O.
- Carbon is oxidised to carbon dioxide, CO_2.

For example, methane burning completely:

$$CH_4 + 2O_2 \rightarrow CO_2 + 2H_2O$$

Carbon dioxide is a greenhouse gas, linked to global warming and climate change. It can be detected in the lab using limewater, which forms a cloudy white suspension of calcium carbonate.

Incomplete combustion

Incomplete combustion happens when there is insufficient oxygen to oxidise the alkanes fully. Hydrogen is still oxidised to water, but **less energy** is released. Other products form in addition to CO_2:
- unburned hydrocarbons
- carbon particulates (seen as soot and smoke)
- carbon monoxide, CO.

A colourless, odourless and toxic gas that reduces the oxygen-carrying capacity of red blood cells. It can kill when present in high enough concentrations.

Sulfur dioxide

Hydrocarbon fuels naturally contain sulfur compounds. Unless these are removed at the refinery before the fuel is used, the sulfur is oxidised to sulfur dioxide during combustion:

$$S + O_2 \rightarrow SO_2$$

Nitrogen oxides, NO_x

Various nitrogen oxides are represented by the general formula, NO_x. These gases:
- are not produced from impurities in the fuel
- are produced by the reaction of nitrogen and oxygen from the air in the high temperatures of engines, e.g. $N_2 + O_2 \rightarrow 2NO$

Acidic oxides

Sulfur dioxide and NO_x form acidic solutions when they dissolve in water. This can be shown using a suitable indicator or a pH meter.

Acid rain

Sulfur dioxide and nitrogen oxides cause **acid rain**. They dissolve in water in clouds to form:
- sulfurous acid, H_2SO_3, which may be oxidised to sulfuric acid, H_2SO_4
- nitrous acid, HNO_2, which may be oxidised to nitric acid, HNO_3.

Acid rain damages the environment. It can:
- ☑ react with stone and metals
- ☑ kill aquatic animal and plant life
- ☑ kill trees and other land plants.

Worked example

Vehicle exhaust systems are fitted with catalytic converters to reduce the release of pollutants.

(a) Name two catalysts used in these devices. **(1 mark)**

Platinum and rhodium.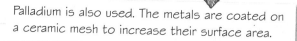

Palladium is also used. The metals are coated on a ceramic mesh to increase their surface area.

(b) Write an equation to show how nitrogen monoxide and carbon monoxide may be removed in a catalytic converter. **(1 mark)**

$$2NO + 2CO \rightarrow N_2 + 2CO_2$$

This is a **redox reaction** (NO is reduced and CO is oxidised). Unburned hydrocarbons may also be oxidised (to carbon dioxide and water).

Now try this

1 Write an equation for the incomplete combustion of ethane to form carbon monoxide, water and a solid product. **(1 mark)**

2 Describe the problems arising from the pollutants carbon monoxide, sulfur dioxide and nitrogen oxides. **(3 marks)**

Alternative fuels

Alternative fuels include biodiesel and bioalcohols derived from renewable sources such as plants.

Fuels

A **fuel** is a substance that can release energy usefully through chemical reactions.

The conventional fuels are **fossil fuels**, fuels formed from the ancient remains of living organisms over millions of years.

The fossil fuels are **non-renewable**:

- Once they have all been used up, they cannot be replaced.

Alternative fuels are **renewable**, and they may be used instead of conventional fuels:

Conventional fuel	Alternative fuel
coal	wood
petrol	bioethanol
diesel	biodiesel
natural gas	biogas

Carbon neutrality

The term **carbon neutral** refers to an activity that has no overall annual carbon emissions to the atmosphere.

The carbon emissions involved are usually of carbon dioxide, but other greenhouse gases such as methane may be included.

Any process that uses fossil fuels in the production or transport of an alternative fuel will reduce the fuel's carbon neutrality.

Renewable energy resources

Alternative fuels are generally renewable energy resources, but not all **renewable energy resources** are fuels. For example, wind, wave, solar and tidal power are not alternative fuels.

Biodiesel

Biodiesel is made from vegetable oils such as those from rapeseed, sunflower or soya.

✗ The 'transesterification' process used to make biodiesel needs a base such as sodium hydroxide, which must be manufactured.

✓ An alcohol is one of the reactants in making biodiesel, but this could be bioethanol.

✓ Biodiesel can be used in unmodified diesel engines, with conventional diesel or on its own.

Reduced dependency on fossil fuels

Countries without large fossil fuel reserves must rely on others for their fuels. Locally produced biofuels reduces this dependency, creates jobs at home and reduces imports.

Bioalcohol

Bioethanol is the most common bioalcohol (alcohol derived from materials produced by living organisms).

It is produced by fermentation of plant sugars:

$$C_6H_{12}O_6 \rightarrow 2CH_3CH_2OH + 2CO_2$$

catalysed by yeast enzymes at about 36 °C

Bioethanol is mixed with petrol for use in ordinary petrol engines, or used on its own in modified engines.

✗ Carbon dioxide is produced during fermentation and during combustion of the fuel.

✓ An equal amount of carbon dioxide is used in plant **photosynthesis** needed to make the sugars.

The use of farmland to grow crops for fuels, rather than to feed people, is also a significant drawback of alternative biofuels.

Worked example

Ethanol can be manufactured by the fermentation of sugars or by the hydration of ethene. Other than cost or energy use, give two advantages of manufacturing ethanol by fermentation, and two disadvantages.   **(4 marks)**

Making ethanol by fermentation uses a renewable raw material and needs less complicated equipment. On the other hand, fermentation is a slower process and bioethanol must be distilled to purify it.

Now try this

1 (a) Explain why bioethanol may be considered to be a carbon neutral fuel.   **(2 marks)**
  (b) State one reason why bioethanol may not actually be carbon neutral.   **(1 mark)**

2 Suggest one environmental reason why alternative fuels may be preferred to fossil fuels.   **(1 mark)**

Alcohols and halogenoalkanes

The halogenoalkanes and the alcohols form homologous series with different functional groups.

Alcohols

The **alcohols** have this functional group:

- the **hydroxyl** group, –OH   Do not confuse this with the hydroxide ion, OH⁻.

They are named according to the number of carbon atoms in the longest chain to which the hydroxyl group is attached.

Their names have the suffix ol. For example:

methanol, CH_3OH ethanol, CH_3CH_2OH

Alcohols with fewer than 11 carbon atoms are usually liquids and the rest are usually solids.

Homologous series

Compounds like alkanes, alkenes, halogenoalkanes and alcohols each belong to different **homologous series**.
A homologous series has these features:

- ✓ the same functional group
- ✓ the same general formula
- ✓ similar chemical properties
- ✓ successive members differ by one –CH_2– group
- ✓ show a trend in physical properties, such as boiling temperature.

For example, the straight chain alkenes have the functional group C=C and the general formula C_nH_{2n}.

They decolourise bromine water and their boiling temperature increases as the number of carbon atoms increases.

Halogenoalkanes

The functional group in a **halogenoalkane** is a halogen atom. Halogenoalkanes are named after this group and the number of carbon atoms in the longest chain to which it is attached.

Locants are used to name position isomers.

1-bromopropane, CH_3CH_2Br 2-bromopropane, $CH_3CHBrCH_3$

Their names have prefixes:

Functional group	Prefix
F	fluoro
Cl	chloro
Br	bromo
I	iodo

The functional groups are named alphabetically if there are two or more different halogens, e.g.

3-bromo-2,3-dichloro-1,1,1-trifluoro-2-iodopropane (this has the lowest sum of locants)

Worked example

Name this alcohol.

3-methylbutan-2-ol

(1 mark)

The longest chain containing the hydroxyl group has four carbon atoms, so the name is based on butane, with the functional group taking priority:
- The hydroxyl group is at position 2.
- The methyl group is at position 3.

If it also had a hydroxyl group on the left hand C atom, it would become 2-methylbutane-1,3-diol.

Now try this

1 Draw the displayed formulae for:
   (a) propan-2-ol, (b) propane-1,3-diol,
   (c) 2-methylpropan-2-ol   **(3 marks)**

2 Draw the displayed formulae for:
   (a) 2-chloropropane,
   (b) 2-bromo-1,2-dichloropropane,
   (c) 2-fluoro-1,3-diiodo-2-methylpropane   **(3 marks)**

Substitution reactions of alkanes

Alkanes can undergo **radical substitution reactions** with halogens to produce halogenoalkanes.

Bond breaking

There are two possibilities when a covalent bond X–Y breaks:

$$X–Y \rightarrow X^+ + {:}Y^- \qquad X–Y \rightarrow X{\cdot} + {\cdot}Y$$

Heterolytic fission – one atom gains both bonding electrons, and ions form

Homolytic fission – each atom keeps a bonding electron, and radicals form

A **radical** (also called a free radical) is a species with an unpaired electron.

The **unpaired electron** is represented as a dot in reaction mechanisms:

- The dot is placed next to the atom whose bond was broken to form the radical.

Substitution reactions

In a **substitution** reaction:

✓ one atom or group is replaced by another atom or group.

For example, methane reacts with chlorine:

$$CH_4 + Cl_2 \rightarrow CH_3Cl + HCl$$

In this reaction, a hydrogen atom in the methane molecule is replaced by a chlorine atom.

Reaction mechanisms

The example above shows you what happens overall, but it does not show the steps involved.

A **reaction mechanism** does show you the steps involved, including:

✓ bonds broken or formed

✓ species (atom, ion, molecule, radical) involved.

The chlorination of methane

The chlorination of methane involves a radical substitution mechanism with **three steps**.

Step name	What is involved	Example
Initiation	Homolytic fission produces chlorine radicals	$Cl_2 \rightarrow {\cdot}Cl + {\cdot}Cl$
Propagation	Reactions between radicals and molecules	$CH_4 + {\cdot}Cl \rightarrow {\cdot}CH_3 + HCl$ ${\cdot}CH_3 + Cl_2 \rightarrow CH_3Cl + {\cdot}Cl$
Termination	Reactions between pairs of radicals to form molecules	${\cdot}CH_3 + {\cdot}Cl \rightarrow CH_3Cl$ ${\cdot}CH_3 + {\cdot}CH_3 \rightarrow C_2H_6$ ${\cdot}Cl + {\cdot}Cl \rightarrow Cl_2$

The Cl–Cl bond is weaker than the C–H bond so is more likely to break when exposed to ultraviolet light or sunlight.

A methyl radical is made in the first reaction. Another chlorine radical is made in the second reaction.

Various by-products are possible, depending on which radicals meet.

Now try this

1 What is a radical and how does it form?   **(2 marks)**
2 Ethane reacts with bromine to produce bromoethane: $CH_3CH_3 + Br_2 \rightarrow CH_3CH_2Br + HBr$
   Write equations to describe the radical substitution mechanism involved.   **(4 marks)**

The mechanism is similar to the one above but an ethyl radical ${\cdot}CH_2CH_3$ forms in the propagation step.

Alkenes and hydrogen halides

Alkenes can undergo **electrophilic addition reactions** with hydrogen halides to produce halogenoalkanes.

Sigma bonds and pi bonds

Sigma bonds (σ bonds) form by end-on overlap between:

> May lead to polar bonds as two different elements are involved.

- ✓ two s-orbitals
- ✓ two p-orbitals
- ✓ an s-orbital and a p-orbital.

Pi bonds (π bonds) form:

- ✓ by sideways overlap between two p-orbitals
- ✓ once a σ bond has formed.

> Pi bonds can only exist between atoms joined by double or triple bonds.

See page 10 for diagrams that describe this.

Bonding in alkenes

The C=C bond in alkenes has one C–C sigma bond and one C–C pi bond (which produces regions of negative charge above and below):

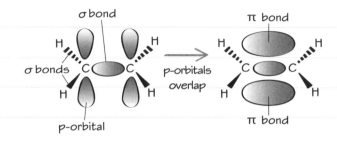

Halogenoalkanes from alkenes

Alkenes react with hydrogen halides to form halogenoalkanes. For example:

$$CH_2=CH_2 + HBr \rightarrow CH_3CH_2Br$$

| ethene | hydrogen bromide | bromoethane |

This is an addition reaction because two reactants produce one product. The mechanism involves an electrophile, $H^{\delta+}$, from the HBr.

Electrophiles

An **electrophile** is a species able to accept a pair of electrons. Electrophiles are attracted to a region of negative charge and include:
- positively charged ions such as H^+ and NO_2^+
- atoms with a partial positive charge, $\delta+$, because they are covalently bonded to a more electronegative atom, for example H in HBr.

Worked example

Outline the mechanism for the reaction between ethene and hydrogen bromide. **(4 marks)**

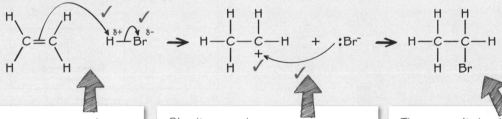

The curly arrow represents the movement of a pair of electrons. Hydrogen bromide has a permanent dipole as bromine is more electronegative than hydrogen. The electrons in the π bond move towards the $H^{\delta+}$ atom.

Simultaneously:
- The H–Br breaks by heterolytic fission, forming H^+ and Br^- ions.
- The π bond breaks and a dative covalent bond forms with the H^+ ion.

The positively charged species formed is a **carbocation**.

The oppositely charged carbocation and bromide ion attract each other and a dative covalent bond forms between them.

The whole mechanism is called **electrophilic addition**.

Now try this

But-2-ene, $CH_3CH=CHCH_3$, reacts with hydrogen bromide to form 2-bromobutane, $CH_3CH_2CHBrCH_3$.
(a) Name the mechanism involved. **(1 mark)**
(b) Outline the mechanism for the reaction. **(4 marks)**

More addition reactions of alkenes

Unsymmetrical alkenes such as propene and but-1-ene undergo electrophilic addition reactions in which two products form, a **major product** and a **minor product** in lower abundance.

Halogenation

Halogenation is a reaction in which a halogen is added to another substance. For example:

$$CH_2=CH_2 + Br_2 \rightarrow CH_2BrCH_2Br$$
  ethene   bromine   1,2-dibromoethane

This is a dihalogenoalkane (alkenes react with hydrogen halides to produce halogenoalkanes).

The dibromo product is colourless, explaining what you see in the bromine test for unsaturation.

Generating electrophiles

The halogenation of alkenes proceeds by an electrophilic addition mechanism.

Unlike hydrogen halides, which are polar, halogen molecules are **non-polar**:

* There are no permanent dipoles in the halogen molecules.

As the halogen molecule approaches the C=C bond in the alkene molecule:

* The electrons in the π bond repel the electrons in the halogen–halogen bond.

* A dipole is induced in the halogen molecule.

After this, the electrophilic addition mechanism is the same as for the reaction of alkenes with hydrogen halides (see previous page).

The reaction with hydrogen proceeds in the same way – a dipole is induced in the hydrogen molecule as it approaches the π bond.

Only one product would form in the reaction between but-1-ene and bromine or hydrogen, as the primary and secondary carbocations both lead to the same product.

Hydrogenation

Hydrogenation is a reaction in which hydrogen is added to another substance. For example:

$$CH_2=CH_2 + H_2 \rightarrow CH_3CH_3$$
  ethene   hydrogen   ethane
  unsaturated   saturated

A nickel **catalyst** is used in hydrogenation.

Hydrogenation is used to make saturated fats from unsaturated vegetable oils for margarine.

Ethene reacts at about 150 °C but vegetable oils are only heated to about 60 °C.

Stability of carbocations

Carbocations are more **stable** if the positive charge is not localised on one carbon atom.

Alkyl groups, such as $-CH_3$ and $-CH_2CH_3$, are electron-releasing.

They reduce the localisation of the positive charge and make the carbocations more stable.

Type of carbocation	General structure	Stability
Primary, 1°	$R_1 - \overset{\underset{\mid}{H}}{\underset{+}{C}} - H$	least stable
Secondary, 2°	$R_1 - \overset{\underset{\mid}{H}}{\underset{+}{C}} - R_2$	
Tertiary, 3°	$R_1 - \overset{\underset{\mid}{R_3}}{\underset{+}{C}} - R_2$	most stable

R_1, R_2 and R_3 represent alkyl groups, which can be identical to or different from each other.

Worked example

Two products form when but-1-ene reacts with hydrogen bromide. Identify the major product formed, and explain your answer.   **(2 marks)**

The major product is 2-bromobutane. It is formed via a secondary carbocation, which is more stable than the primary carbocation.

$$CH_3 - CH_2 - CH_2 - \overset{+}{C}H_2 \quad \text{——1° carbocation}$$

$$CH_3 - CH_2 - \overset{+}{C}H - CH_3 \quad \text{——2° carbocation}$$

Now try this

Propene, $CH_3CH=CH_2$, reacts with hydrogen bromide.
(a) Name the major product and the minor product.   **(2 marks)**
(b) Outline the mechanism for the reaction that produces the major product.   **(4 marks)**

Exam skills 5

This exam-style question uses knowledge and skills you have already revised. Have a look at pages 50–52 for a reminder about **substitution reactions of alkanes** and **addition reactions of alkenes**.

Worked example

Bromoethane, CH_3CH_2Br, can be made from ethane by a free radical substitution reaction. It can also be made from ethene by an electrophilic addition reaction.

$$C_2H_6 + Br_2 \rightarrow CH_3CH_2Br + HBr$$
$$C_2H_4 + HBr \rightarrow CH_3CH_2Br$$

The two types of reaction are given in this question, but be prepared to state them.

(a) (i) State the reagent and condition needed to make bromoethane from ethane.   **(2 marks)**

Bromine and ultraviolet light.

You could state a temperature between 450 °C and 1000 °C instead of UV light.

(ii) The initiation step is: $Br_2 \rightarrow \bullet Br + \bullet Br$

Explain why this is more likely than
$C_2H_6 \rightarrow \bullet CH_2CH_3 + \bullet H$ **(2 marks)**

The Br–Br bond is weaker than the C–H bond, so it is more easily broken.

Note that the dot representing the unpaired electron is placed next to the atom with the broken covalent bond.

You could also answer in terms of bond enthalpies: the Br–Br bond enthalpy is less than the C–H bond enthalpy.

(iii) Write equations for the propagation step.   **(2 marks)**

$C_2H_6 + \bullet Br \rightarrow \bullet CH_2CH_3 + HBr$
$\bullet CH_2CH_3 + Br_2 \rightarrow CH_3CH_2Br + \bullet Br$

Remember to show a halogen radical reacting in your first equation and a halogen radical being formed in your second equation. Check that the numbers of dots are the same on both sides.

(iv) Explain, with the help of an equation, why butane also forms.   **(2 marks)**

Two ethyl radicals may react together to form butane in an termination step:
$2 \bullet CH_2CH_3 \rightarrow C_4H_{10}$

Other termination steps are possible, such as:
$\bullet CH_2CH_3 + \bullet Br \rightarrow CH_3CH_2Br$
However, this question clearly asks you to explain the formation of butane.

(b) Give the mechanism for the reaction between ethene and hydrogen bromide.   **(3 marks)**

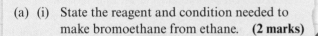

Make sure you practise drawing reaction mechanisms using curly arrows. Remember that each one represents the movement of a pair of electrons:
• Start the arrow on a covalent bond or next to a lone pair of electrons.
• Point the head towards the destination of the pair of electrons.
Remember to show the structure of the intermediate carbocation clearly, and include the + and − charges.

(c) Compare, without calculation, the atom economies of the two processes for making bromoethane.   **(1 mark)**

The reaction between ethene and hydrogen bromide has the higher atom economy. It is 100% because there is only one product.

Maths skills If a calculation is not needed, the answer should be clear to you using your knowledge and understanding of chemistry.

It is important to state that its atom economy is 100%, not just higher than the atom economy of the other process.

Alkenes and alcohols

Alkenes can undergo hydration reactions to produce alcohols, and alcohols can undergo dehydration reactions to produce alkenes.

Alcohols from alkenes

Alkenes react with steam in the presence of a phosphoric acid catalyst. For example:

$$CH_2=CH_2 + H_2O \rightarrow CH_3CH_2OH$$

　　ethene　　steam　|　ethanol
　　　　　300°C, 7 MPa pressure

This reaction proceeds via an electrophilic addition mechanism. It is an example of a **hydration** reaction.

Do not confuse this with hydrogenation!

Alkenes from alcohols

Alcohols undergo elimination reactions to form alkenes. For example:

$$CH_3CH_2OH \rightarrow CH_2=CH_2 + H_2O$$

　　ethanol　|　ethene　　steam
Al_2O_3 catalyst at 300°C, or 180°C with concentrated phosphoric acid as a catalyst

This reaction proceeds via an elimination mechanism similar to the one that occurs in halogenoalkanes (see page 59). It is an example of a **dehydration** reaction.

Worked example

Use your understanding of electrophilic addition reactions in alkenes to suggest the mechanism for the reaction between ethene and steam, in the presence of H^+ ions from an acid catalyst. **(4 marks)**

Hydrogen ions can act as electrophiles because they can accept a pair of electrons.

A carbocation forms, which then forms a dative covalent bond with a water molecule.

An unstable intermediate forms.

An O–H bond breaks, releasing a hydrogen ion.

Notice that an H^+ ion is used at the start, and one is produced at the end. It acts as a catalyst.

Formation of diols

Alkenes can react with an **oxidising agent** to produce alcohols containing two hydroxyl groups. These alcohols are called **diols**.

The oxidising agent is potassium manganate(VII), $KMnO_4$, acidified with dilute sulfuric acid.

It is represented as [O] in chemical equations.

For example:

$$CH_2=CH_2 + [O] + H_2O \rightarrow CH_2OH–CH_2OH$$

　ethene　　　　　　　　ethane-1,2-diol

A colour change from purple to colourless is observed, so this reaction can be used as a test for unsaturation. As with the bromine or bromine water test, no change is observed with alkanes.

Two organic products

Alcohols can be classified as primary, secondary or tertiary alcohols (see pages 57, 60 and 61).

Only one organic product is possible from the dehydration of a primary alcohol. For example:

$$CH_3CH_2CH_2CH_2OH \rightarrow CH_3CH_2CH=CH_2 + H_2O$$

　butan-1-ol　　　　　　but-1-ene　　　steam

The C=C bond can only form in one place here.

This is also true for the dehydration of propan-2-ol, but in other secondary alcohols the C=C bond may be able to form in one of two places.

Two organic products are possible from a secondary alcohol such as butan-2-ol:

✓ but-1-ene, $CH_3CH_2CH=CH_2$
✓ but-2-ene, $CH_3CH=CHCH_3$

Now try this

Write the equation, and name the organic product, for the reaction between but-2-ene and:

(a) steam **(2 marks)**

(b) acidified potassium manganate(VII). **(2 marks)**

Addition polymerisation

Alkenes can act as monomers, forming addition polymers, such as poly(ethene) from ethene.

Monomers

Alkenes can undergo addition reactions because they have C=C bonds.

They can react together, with the C=C bonds joining end to end to produce very large molecules called **polymers**.

These polymers are **addition polymers**, since they were formed by addition reactions.

The alkenes that produced the polymers are called **monomers**. In general:

monomer molecules   polymer molecule

Naming polymers

Addition polymers are named for the monomer from which they are produced. For example:

Monomer	Polymer
ethene	poly(ethene)
propene	poly(propene)
phenylethene	poly(phenylethene)
chloroethene	poly(chloroethene)
tetrafluoroethene	poly(tetrafluoroethene)

polystyrene

—PVC

PTFE, Teflon®

Note that the last two monomers are not alkenes but they do contain C=C bonds.

Repeat units from monomers

The number of monomer molecules making a particular polymer molecule varies, so a **repeat unit** is used to represent the polymer.

You work out a repeat unit from a monomer by:

• converting the C=C bond to a –C–C– bond

• showing the bond angles at 90°.

In general:

```
  W       X           ┌  W   X  ┐
   \     /            │   │   │  │
    C = C         ─┤ ─C ─ C─ ├─
   /     \            │   │   │  │
  Y       Z           └  Y   Z  ┘
   monomer         repeat unit of polymer
                           formed
```

Monomers from polymers

This is a section of a Perspex® molecule:

```
    H    CH₃   H    CH₃   H    CH₃
    │    │     │    │     │    │
 ─ C ─ C ─ C ─ C ─ C ─ C ─
    │    ‖     │    ‖     │    ‖
    H    C     H    C     H    C
        O⟋ ⟍O     O⟋ ⟍O     O⟋ ⟍O
              │          │          │
             CH₃       CH₃        CH₃
```

Here is its monomer: C=C instead of C–C

```
  H        CH₃
   \      /
    C == C          identify where
   /      \         the repeat
  H        C ─ O    happens
           ‖     \
           O      CH₃
```

Write an equation to show the formation of poly(propene) from its monomer.   **(2 marks)**

```
   H     CH₃              ┌  H    CH₃ ┐
    \    /                │   │    │   │
 n   C = C      ──→    ─┤ ─C ─ C─ ├─
    /    \                │   │    │   │
   H     H               └  H    H   ┘ₙ
```

Propene is the monomer for poly(propene).

The n in front of its displayed formula represents the number of monomer molecules reacting.

The repeat unit is drawn inside brackets:

• The brackets can be square or curved.

• The bonds linking to the next repeat unit pass through the brackets.

• n is shown as a subscript at the bottom right of the repeat unit.

A section of a polymer is shown on the right:
(a) Draw the displayed formula for its monomer.   (1 mark)
(b) State the name of the monomer.   (1 mark)
(c) Draw the repeat unit for the polymer.   (1 mark)
(d) Name the type of polymer represented.   (1 mark)

```
   H   H   H   H   H   H
   │   │   │   │   │   │
 ─ C ─ C ─ C ─ C ─ C ─ C ─
   │   │   │   │   │   │
   H   Cl  H   Cl  H   Cl
```

Polymer waste

Waste polymers can be sorted into different types for recycling, for incineration to release energy or for use as a chemical feedstock for cracking.

Sorting polymers

Different polymers have different chemical compositions and properties, so waste polymers must be sorted before recycling.

This can be done by:

- hand ——— labour intensive
- machine ——— Polymers have different densities, so they can be separated by flotation in a liquid.

Biodegradable polymers

Addition polymers are not biodegradable:
- 👎 They take a long time to decompose in landfill.
- 👎 They cause a litter nuisance.

Chemists are developing biodegradable polymers, including ways to make addition polymers like poly(ethene) biodegradable:

- Starch is added during manufacture.
- Bacteria can break down the starch if the polymer gets wet.
- The polymer breaks down into tiny pieces.

Life cycle of a polymer

The environmental impact of a polymer can be assessed by a life cycle analysis. This involves making measurements in its **life cycle** during:

- manufacture, including obtaining and processing the crude oil
- distribution and end use of finished products, such as clothes, packaging and TVs
- possible re-use, recycling in various ways and its final disposal.

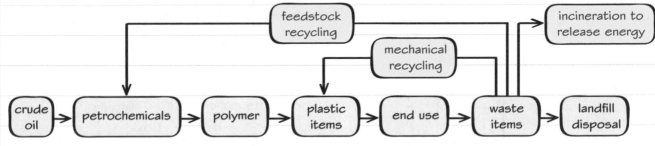

Feedstock recycling

Poly(propene) and similar polymers can be decomposed in a similar to way to cracking, producing a fraction for use in making new polymers.

Mechanical recycling

Some polymers, such as PET (used in plastic bottles) can be shredded, melted and made into new objects such as fleeces and carpets.

Incineration

Polymers can be burned at high temperatures, releasing energy that can be used to heat buildings and to generate electricity.

Worked example

Poly(chloroethene) (or PVC), can be disposed of by incineration. Use your knowledge of the composition of this polymer to suggest a toxic gaseous pollutant that could be released, and how it might be removed.   **(2 marks)**

Hydrogen chloride could be produced because the PVC contains chlorine and hydrogen. It could be absorbed by passing the waste gases through sodium hydroxide solution.

Now try this

1 State three reasons why waste polymers may be sorted into different types.   **(3 marks)**
2 Explain what is meant by feedstock recycling.   **(2 marks)**
3 Give an example of a toxic waste gas that may be produced by the incineration of waste polymers at high temperatures.   **(1 mark)**

Alcohols from halogenoalkanes

Primary halogenoalkanes undergo **nucleophilic substitution** with OH⁻ ions to produce primary alcohols.

Primary halogenoalkanes

In halogenoalkanes, a hydrogen atom has been replaced by a halogen atom (see page 49).

A **primary halogenoalkane** is one where:

✓ The carbon atom to which the halogen atom is attached is directly attached to just one other carbon atom.

1-bromopropane is a primary halogenoalkane.

```
    H   H   H
    |   |   |
H—C—C—C—Br
    |   |   |     ╲only directly attached
    H   H   H      to one C atom
```

Primary alcohols

Alcohols contain the hydroxyl group, –OH as a functional group (see page 49).

A **primary alcohol** is one where:

✓ The carbon atom to which the hydroxyl group is attached is directly attached to just one other carbon atom.

Propan-1-ol is a primary alcohol.

```
    H   H   H
    |   |   |
H—C—C—C—OH
    |   |   |     ╲only directly attached
    H   H   H      to one C atom
```

Substitution reactions

Primary halogenoalkanes react with potassium hydroxide solution, under reflux conditions, to produce primary alcohols. For example:

$$CH_3CH_2CH_2Br + OH^- \rightarrow CH_3CH_2CH_2OH + Br^-$$
   1-bromopropane   propan-1-ol

This an example of a **substitution reaction**:

- One atom or group is exchanged by another atom or group.

In this case, the halogen atom is replaced by a hydroxyl group and leaves as a halide ion.

Reflux

Reflux is used to heat a reaction mixture for a long time without losing any liquid:
1. Liquid boils and its vapours rise.
2. The vapours are cooled and condense.
3. Condensate returns to the flask.

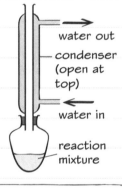

water out
condenser (open at top)
water in
reaction mixture

Worked example

Show the mechanism for the reaction between 1-bromopropane and potassium hydroxide solution under reflux conditions. **(2 marks)**

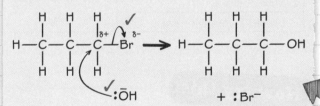

The OH⁻ ion acts as a **nucleophile**, a species that can donate a lone pair of electrons to form a dative covalent bond. Nucleophiles can be:
- negatively charged ions, e.g. OH⁻ or CN⁻
- molecules with lone pairs, e.g. H_2O or NH_3

The C–Br bond is polar because bromine is more electronegative than carbon, so:
- The C atom has a partial positive charge
- It can be attacked by nucleophiles.

A dative covalent bond forms between the electron-deficient C atom and the O atom in the OH⁻ ion. At the same time, the bonding pair of electrons in the C–Br bond moves to the Br atom. The bond breaks and a Br⁻ ion is released.

Now try this

1 What is a nucleophile? **(1 mark)**

2 Outline the nucleophilic substitution mechanism for the reaction between 1-chloroethane and potassium hydroxide solution under reflux conditions. Name the organic product. **(3 marks)**

Reactivity of halogenoalkanes

The reactions between halogenoalkanes and water can be followed using silver nitrate solution in ethanol.

Hydrolysis of halogenoalkanes

Halogenoalkanes react with water to form alcohols. In general:

an alkyl group,
e.g. CH_3CH_2

$$RX + H_2O \rightarrow ROH + H^+ + X^-$$

halogen atom   halide ion

For example:

$$CH_3CH_2Br\ H_2O \rightarrow CH_3CH_2OH + H^+ + Br^-$$
1-bromoethane   ethanol

These are **hydrolysis reactions**:

* reactions in which a molecule of water is reacted with a substance.

They proceed by a nucleophilic substitution mechanism in which water is the nucleophile.

Secondary and tertiary halogenoalkanes

2-bromobutane is a **secondary** halogenoalkane.

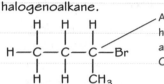

Atom attached to halogen is directly attached to two C atoms.

2-bromo-2-methylpropane is a **tertiary** halogenoalkane.

Atom attached to halogen is directly attached to three C atoms.

Studying hydrolysis

The hydrolysis of halogenoalkanes is studied using silver nitrate solution mixed with ethanol.

Ethanol is used as the solvent for the mixture because it prevents two layers forming.

The silver ions react with the halide ions released in the reaction, forming a precipitate:

* AgCl is white.
* AgBr is cream coloured.
* AgI is yellow.

If the reaction conditions are constant (volumes, concentrations and temperature), the faster the reaction, the faster the precipitate appears.

Ag^+ and F^- ions do not form a precipitate, so cannot be detected this way (see page 31).

Different halogens

For a given carbon chain, the reactivity of halogenoalkanes increases Cl < Br < I. For example:

Halogenoalkane	Rate of hydrolysis
1-chlorobutane	slowest
1-bromobutane	↓
1-iodobutane	fastest

Different structures

For a given halogen, the reactivity increases: primary < secondary < tertiary. For example:

Halogenoalkane	Rate of hydrolysis
1-bromobutane (1°)	slowest
2-bromobutane (2°)	↓
2-bromo-2-methylpropane (3°)	fastest

Worked example

Explain the trend in the reactivity of primary chloroalkanes, bromoalkanes and iodoalkanes. **(3 marks)**

The carbon–halogen bond is broken in the reaction. The C–Cl bond is the strongest and the C–I bond is the weakest. The lower the bond enthalpy, the more easily the bond is broken and the more reactive the halogenoalkane.

Now try this

Compared to 1-chlorobutane, 1-fluorobutane is expected to:

☐ **A** react more quickly because fluorine is more reactive than chlorine.

☐ **B** react more quickly because the C–F bond is weaker than the C–Cl bond.

☐ **C** react more slowly because 1-fluorobutane is less soluble than 1-chlorobutane.

☐ **D** react more slowly because the C–F bond is stronger than the C–Cl bond. **(1 mark)**

More halogenoalkane reactions

Halogenoalkanes can undergo substitution or elimination reactions depending on the conditions.

Making nitriles

Halogenoalkanes react under reflux with potassium cyanide, KCN, dissolved in ethanol to produce **nitriles**. For example:

$$CH_3CH_2Br + CN^- \rightarrow CH_3CH_2CN + Br^-$$
1-bromoethane   propanenitrile

The product contains one more C atom than the organic reactant – a useful way to increase the chain length.

These reactions proceed by a nucleophilic substitution mechanism in which the CN^- ion is the nucleophile. It is like the mechanism on page 57 but the nucleophile is $:\overset{..}{C}N$ rather than $:\overset{..}{O}H$.

Nitriles contain the **nitrile group**, $-C\equiv N$.

They are named according to the number of carbon atoms in the longest chain containing the nitrile group:
- CH_3CH_2CN is propanenitrile not ethanenitrile.

Making amines

Primary halogenoalkanes react with ammonia to make **primary amines** when heated in a sealed tube. For example:

$$CH_3CH_2CH_2Br + 2NH_3 \rightarrow CH_3CH_2CH_2NH_2 + NH_4Br$$
1-bromopropane   propylamine

Excess ammonia is used to reduce the chance of the primary amine product reacting with the halogenoalkane.

Amines contain the **amino group**, $-NH_2$.

You can name primary amines in three ways. For example:

1° amine	CH_3NH_2	$CH_3CH_2NH_2$
Names	methylamine	ethylamine	commonly used
	aminomethane	aminoethane	
	methanamine	ethanamine	IUPAC recommended

Nucleophilic substitution

Here is the nucleophilic substitution reaction mechanism for reaction between 1-bromopropane and excess ammonia. Note that two ammonia molecules are involved (not the same one twice).

Ammonia as a **nucleophile** – it donates a pair of electrons to form a dative covalent bond.

Ammonia as a **base** – it accepts a **hydrogen ion**, H^+ (you can also say it accepts a **proton**).

When NH_3 accepts the H^+ it becomes an NH_4^+ ion. With the Br^- ion it is shown as NH_4Br.

Worked example

(a) Write an equation for the reaction between 2-bromopropane and ethanolic potassium hydroxide. Name the organic product. **(2 marks)**

$$CH_3CHBrCH_3 + KOH \rightarrow CH_3CH=CH_2 + H_2O + KBr$$

The organic product is propene.

When potassium hydroxide is dissolved in ethanol, the OH^- ion acts as a base rather than as a nucleophile and an **alkene** is formed.

In the reaction:
- An H atom and a Br atom are removed.
- They are not replaced by other atoms.

This type of reaction is called **elimination**.

(b) State the type of reaction involved. **(1 mark)**

It is an elimination reaction.

Top tip! You do not need to know the mechanism for this reaction.

Now try this

Outline the reaction mechanism for the reaction between chloroethane and excess ammonia. Name the organic product.

(5 marks)

Oxidation of alcohols

Alcohols react with oxygen in the air to form carbon dioxide and water. This combustion reaction is an example of an oxidation reaction, but alcohols can also be oxidised in other ways.

Aldehydes and ketones

These contain the **carbonyl group**, >C=O. It is at the end of a carbon chain in **aldehydes**, but not at the end in **ketones**.

They are named by the longest carbon chain with the carbonyl group, ending in:

✓ -al for aldehydes   ✓ -one for ketones.

Aldehyde	Ketone
Propanal	Propanone
CH_3CH_2CHO	CH_3COCH_3

Carboxylic acids

The **carboxylic acids** have the carboxyl group, –COOH.

They react with carbonates and turn universal indicator orange or red, just as other acids do.

They are named by the longest carbon chain with the carboxyl group, ending in -anoic acid.

For example:

propanoic acid   2-methylpropanoic acid

Making aldehydes and ketones

Potassium dichromate(VI), $K_2Cr_2O_7$, acidified with sulfuric acid, is an oxidising agent.

When alcohols are heated with this oxidising agent under **distillation**:

- Primary alcohols produce aldehydes.
- Secondary alcohols produce ketones.

For example:

$$CH_3CH_2CH_2OH + [O] \rightarrow CH_3CH_2CHO + H_2O$$
propan-1-ol   propanal

$$CH_3CH(OH)CH_3 + [O] \rightarrow CH_3COCH_3 + H_2O$$
propan-2-ol   propanone

When a primary alcohol is heated with this oxidising agent under **reflux**, the aldehyde formed is further oxidised to a carboxylic acid:

$$CH_3CH_2CHO + [O] \rightarrow CH_3CH_2COOH$$
propanal   propanoic acid

Practical skills   Distillation

If you use reflux, the reactants and products remain mixed together (see page 57).

If you use distillation instead, the product leaves the reaction mixture as it forms.

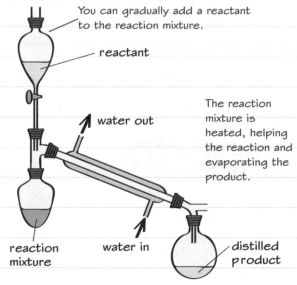

You can gradually add a reactant to the reaction mixture.

reactant

water out

The reaction mixture is heated, helping the reaction and evaporating the product.

reaction mixture   water in   distilled product

(see page 57).

Worked example

Describe a simple laboratory test to distinguish between an aldehyde and a ketone.   **(3 marks)**

Add a few drops of Benedict's solution to each compound, then warm the mixture. There will be a colour change from blue to red with the aldehyde, but no visible change with the ketone.

You could use Fehling's solution instead, and you would observe the same changes.

Now try this

Write equations to show the oxidation of alcohols with acidified potassium dichromate(VI):
(a) butan-1-ol under reflux, and under distillation.   **(3 marks)**
(b) butan-2-ol under distillation.   **(1 mark)**

Halogenoalkanes from alcohols

Alcohols are **halogenated** to form halogenoalkanes in different ways, depending upon the halogen involved.

Chloroalkanes from alcohols

Phosphorus(V) chloride, PCl_5, is a white solid at room temperature. It reacts vigorously with alcohols to produce chloroalkanes. In general:

$$ROH + PCl_5 \rightarrow RCl + POCl_3 + HCl$$

No heat is needed and white fumes of hydrogen chloride are produced. For example:

$$\underset{\text{ethanol}}{CH_3CH_2OH} + PCl_5 \rightarrow \underset{\text{chloroethane}}{CH_3CH_2Cl} + POCl_3 + HCl$$

Tertiary alcohols

Tertiary alcohols (not primary or secondary alcohols) react quickly with concentrated hydrochloric acid at room temperature.

For example:

$$\underset{\text{2-methylpropan-2-ol}}{(CH_3)_3COH} + HCl \rightarrow \underset{\text{2-chloro-2-methylpropane}}{(CH_3)_3CCl} + H_2O$$

Bromoalkanes from alcohols

A mixture of potassium bromide in 50% concentrated sulfuric acid reacts with alcohols to produce bromoalkanes.

Hydrogen bromide is produced:

$$KBr + H_2SO_4 \rightarrow KHSO_4 + HBr$$

$$2KBr + H_2SO_4 \rightarrow K_2SO_4 + 2HBr$$

The hydrogen bromide reacts with the alcohol, e.g.

$$\underset{\text{propan-1-ol}}{CH_3CH_2CH_2OH} + HBr \rightarrow \underset{\text{1-bromopropane}}{CH_3CH_2CH_2Br} + H_2O$$

Iodoalkanes from alcohols

A mixture of red phosphorus and iodine reacts with alcohols, under reflux conditions, to produce iodoalkanes. Phosphorus(III) iodide is produced:

$$2P + 3I_2 \rightarrow 2PI_3$$

This reacts with the alcohol. For example:

$$\underset{\text{ethanol}}{3CH_3CH_2OH} + PI_3 \rightarrow \underset{\text{iodoethane}}{3CH_3CH_2I} + H_3PO_3$$

Solvent extraction

A **separating funnel** is used to separate two **immiscible** liquids (liquids that do not mix).

Some reactions may also be carried out in it. For example, the reaction between 2-methylpropan-2-ol and concentrated hydrochloric acid:

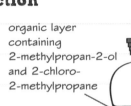

organic layer containing 2-methylpropan-2-ol and 2-chloro-2-methylpropane

aqueous layer containing water and unreacted hydrochloric acid

Drying with an anhydrous salt

In the preparation of 2-chloro-2-methylpropane, the lower layer is run off from the separating funnel and discarded.

The upper layer containing the product is dried to remove any water present. You:

- ✓ run it off into a clean flask
- ✓ add anhydrous sodium sulfate, Na_2SO_4
- ✓ shake the mixture, then **decant** it into another flask.

The dried organic layer can then be **distilled** to separate 2-chloro-2-methylpropane from unreacted 2-methylpropan-2-ol.

Worked example

Describe how you could determine the purity of a liquid.   **(3 marks)**

Measure the boiling temperature of the liquid using distillation apparatus. The closer this is to the accepted temperature, the purer the liquid.

Different compounds often have different boiling temperatures. For example:

- 2-chloro-2-methylpropane boils at 50.6 °C.
- 2-methylpropan-2-ol boils at 82.2 °C.
- Water boils at 100 °C.

Now try this

Name the reactants needed to chlorinate, brominate or iodinate butan-1-ol.   **(3 marks)**

Exam skills 6

This exam-style question uses knowledge and skills you have already revised. Have a look at pages 60 and 61 for a reminder about **oxidation of alcohols** and **halogenoalkanes from alcohols**.

Worked example

(a) Ethanal can be made from ethanol.
  (i) State the names or formulae of the two substances needed to make ethanal from ethanol. **(2 marks)**

Potassium dichromate(VI), acidified with dilute sulfuric acid.

  (ii) Draw a diagram to show the laboratory apparatus needed to make ethanal from ethanol and to collect the ethanal. **(2 marks)**

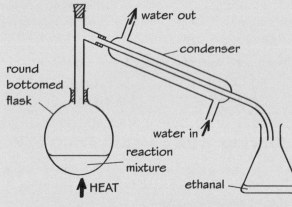

round bottomed flask

water out

condenser

water in

reaction mixture

↑ HEAT

ethanal

  (iii) Describe what would be seen when ethanol and ethanal are warmed separately with Tollens' reagent. **(1 mark)**

There would be no visible change with ethanol but a silver mirror would form with ethanal.

(b) Phosphorus(V) chloride, PCl_5, reacts with ethanol.
  (i) Describe what would be seen during the reaction. **(1 mark)**

Steamy fumes are produced.

  (ii) Write an equation for the reaction. **(1 mark)**

$CH_3CH_2OH + PCl_5 \rightarrow CH_3CH_2Cl + POCl_3 + HCl$

(c) A mixture of ethanol and water can be distilled to separate some of the ethanol. Name a suitable drying agent to absorb remaining water in the distilled ethanol and describe how you would produce dry ethanol using it. **(2 marks)**

Use anhydrous calcium oxide. Mix it with the distilled ethanol, then filter the mixture to remove the solid.

'Potassium dichromate' would not be enough – you need to include the oxidation number of chromium in $K_2Cr_2O_7$. Similarly, 'acid' would not be enough for H_2SO_4 – you need to state its name or formula.

🧪 **Practical skills**   You need to draw apparatus for heating under distillation conditions. If you showed apparatus for reflux instead, further oxidation to ethanoic acid would occur if this apparatus was used.

Make sure you show that heat is needed (a labelled arrow is enough). Take care that your drawing shows:
• the still head sealed so that the vapours could not escape without entering the condenser
• the condenser open at one end so that the apparatus would not explode.

Make sure you can recall the expected observations when experiments are carried out.

The answer describes the expected observations for both compounds.

Command word: Describe

If a question asks you to **describe** something, it means you need to:
• give an account of something
• link statements if necessary.

You do not need to:
• include a justification or reason.

The organic reactant and product are shown using structural formulae rather than molecular formulae.

Ethanol and water can be separated from each other by distillation because they have different boiling points.

Anhydrous sodium sulfate or anhydrous magnesium sulfate could be used instead. You could decant the mixture instead of filtering it (the solid would stay behind).

Structures from mass spectra

It is possible to suggest the structure of a molecule from its mass spectrum.

The molecular ion peak

The peak with the highest m/z ratio represents the **molecular ion**:

- It is the peak furthest to the right.
- Its m/z ratio is equal to the M_r.

Ethanol, CH_3CH_2OH, produces a molecular ion peak M at $m/z = 46.0$, so its M_r is 46.0 too.

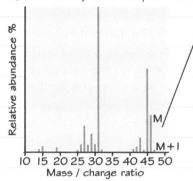

Isotopes such as ^{13}C may cause a very small peak at M+1, which can be ignored when finding M_r.

Fragmentation

Bonds can break in the molecular ion, forming two species:

- a positive ion
- a neutral species, usually a radical.

In a compound X–Y, there are two possibilities:

$$(X–Y)^+ \rightarrow \bullet X + Y^+ \quad \text{and} \quad (X–Y)^+ \rightarrow X^+ + \bullet Y$$

Only a charged fragment will reach the detector, but as both X^+ and Y^+ can be produced, both will appear on a mass spectrum. For example:

```
 H   H   H              H           H   H
 |   |   |              |           |   |
H—C—|—C—|—C—H     H—C+    + C—C—H
 |   |   |              |           |   |
 H   H   H              H           H   H
 |                                      
C–C bond breaks    CH₃⁺ fragment   CH₃CH₂⁺ fragment
in propane         with m/z = 15   with m/z = 29
```

The mass spectrometer

The **mass spectrometer** is a machine that measures the masses of atoms and molecules.

It produces **positive ions** by bombarding the sample with high-energy electrons. In general:

$$X(g) + e^- \rightarrow X^+(g) + 2e^-$$

The positive ions in the mass spectrometer are deflected by a magnetic field according to their mass/charge ratio, m/z (see page 2).

A **mass spectrum** is a chart with:

- relative abundance on the vertical axis
- m/z ratio on the horizontal axis.

Some typical fragments

Here are some fragments and their m/z ratios:

m/z	Fragments		
15	CH_3^+		
29	CHO^+	$CH_3CH_2^+$	
31	CH_3O^+	CH_2OH^+	
43	CH_3CO^+	$CH_3CH_2CH_2^+$	$(CH_3)_2CH^+$
57	$CH_3CH_2CO^+$		
77	$C_6H_5^+$ (from a benzene ring)		

Arenes, compounds containing benzene rings, are studied in Topic 18 at A level.

Other fragments and m/z ratios are possible.

Worked example

The mass spectrum of butane, $CH_3CH_2CH_2CH_3$, has peaks at $m/z = 15$ and $m/z = 43$.
Write equations to show the production of fragments that could produce these peaks. **(2 marks)**

$$(CH_3CH_2CH_2CH_3)^+ \rightarrow CH_3^+ + \bullet CH_2CH_2CH_3$$
$$(CH_3CH_2CH_2CH_3)^+ \rightarrow CH_3CH_2CH_2^+ + \bullet CH_3$$

The relative formula mass of butane is 58.0, so it will have a molecular ion peak at $m/z = 58$.

The peak at $m/z = 15$ could be due to CH_3^+ and the one at $m/z = 43$ to $CH_3CH_2CH_2^+$.

They form when a C–C bond breaks.

Make sure you remember to include the + charge on the molecular ion and on the fragment that is detected.

Now try this

Refer to the mass spectrum of ethanol at the top of this page.
(a) Suggest the identities of the fragments with $m/z = 15$, $m/z = 29$ and $m/z = 31$. **(3 marks)**
(b) For each fragment identified in part (a), write an equation to show its production. **(3 marks)**

Infrared spectroscopy

Infrared spectroscopy is used to identify the presence, or absence, of different functional groups.

Bond vibrations

Covalent bonds vibrate continually, including **stretching** about a mean position.

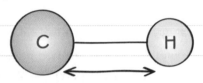

The frequency at which a bond vibrates depends on a number of factors including:
- its bond length
- its bond enthalpy (bond strength)
- the masses of the two bonded atoms.

For example, the C–H group vibrates at about 8.7×10^{13} Hz, which is in the **infrared** range.

Absorbing infrared radiation

A bond can absorb infrared radiation if:
- the bond is polar, and
- the frequency matches the frequency of the bond vibrations.

This is the basis of **infrared spectroscopy**:
- Infrared radiation is passed through a sample.
- Absorptions due to functional groups appear as 'peaks' in the infrared spectrum.

An infrared spectrum is a chart with:
- % transmittance on the vertical axis
- wavenumber in cm^{-1} on the horizontal axis.

Wavenumber, the number of waves in 1 cm, is used instead of frequency.

For example, 8.7×10^{13} Hz is 2900 cm^{-1}.

Worked example

The infrared spectrum of a compound with the molecular formula $C_4H_6O_2$ is shown below.

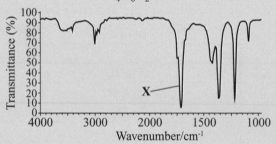

(a) Identify the functional group that could be responsible for the peak labelled X. **(1 mark)**

It could be a C=O group in a ketone or carboxylic acid, as the peak is at about 1710 cm^{-1}.

(b) Suggest the type of compound that produced the spectrum. Explain your answer. **(2 marks)**

It could be a ketone because a carboxylic acid should also produce a peak at 3300–500 cm^{-1} due to its O–H bond and this is missing here.

Wavenumbers for bond stretching

Infrared absorption wavenumbers are in the Data Book **given to you** in the examination.

Functional group	Found in	Wavenumber/cm^{-1}
C–H	alkanes	2962–2853
	alkenes	3095–3010
	aldehydes	2900–2820
		2775–2700
C=C	alkenes	1669–1645
C=O	aldehydes	1740–1720
	ketones	1720–1700
	carboxylic acids	1725–1700
O–H	alcohols	3750–3200
	carboxylic acids	3300–2500
N–H	amines	3500–3300

Causes broad peaks due to hydrogen bonding.

If the peak was 1740–1720 cm^{-1}, the compound could be an aldehyde (for the same reasons).

Now try this

The infrared spectrum of a compound with the molecular formula C_3H_8O is shown on the right.

(a) Identify the functional groups that could be responsible for peaks A and B. **(2 marks)**

(b) Suggest the name and structural formula of the compound involved. Explain your answer. **(3 marks)**

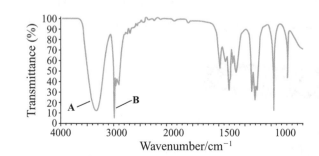

Enthalpy changes

Enthalpy change, ΔH, is the heat energy change of a system measured at constant pressure. In a chemical reaction, the system is the reaction mixture, and the surroundings are everything else.

Exothermic reactions

In an **exothermic** reaction:
- There is an overall transfer of energy from the system to the surroundings.
- The enthalpy change, ΔH, is negative.

The combustion of carbon is exothermic, for example. Here is its **enthalpy level diagram**:

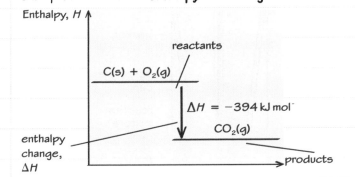

Endothermic reactions

In an **endothermic** reaction:
- There is an overall transfer of energy to the system from the surroundings.
- The enthalpy change, ΔH, is positive.

The reaction between carbon and carbon dioxide is endothermic:

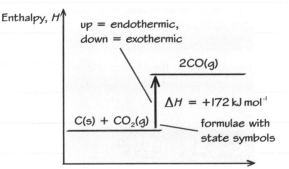

Standard conditions

Enthalpy changes may be compared by referring to **standard conditions**:
- 100 kPa pressure (1×10^5 Pa)
- a stated temperature, usually 298 K (25 °C)
- solutions, if present, at 1 mol dm^{-3}.

Enthalpy changes under standard conditions are **standard enthalpy changes**, $\Delta H^{\ominus}_{298}$ or $\Delta H^{\ominus}$.

The **standard state** of a substance is its state (solid, liquid or gas) under standard conditions.

Recognising energy transfer

Exothermic reactions transfer energy to the surroundings, and the temperature increases.

Endothermic reactions transfer energy from the surroundings:
- in reactions in solution, the surroundings include the solvent, so the temperature decreases
- in reactions such as electrolysis or thermal decomposition of solids, a constant input of energy is needed to maintain it.

Worked example

(a) Define the term standard enthalpy change of formation, $\Delta H^{\ominus}_f$. **(3 marks)**

The enthalpy change measured at 100 kPa and a specified temperature, usually 298 K, when one mole of a substance is formed from its elements in their standard states.

(b) Write an equation to represent the standard enthalpy change of formation of ethanol.
(1 mark)

$2C(s) + 3H_2(g) + \frac{1}{2}O_2(g) \rightarrow CH_3CH_2OH(l)$

The formation of a substance from its elements is just one example of a chemical reaction. In general, the **standard enthalpy change of reaction** has the symbol $\Delta H^{\ominus}_r$.

You assume that all the reactants and products are in their standard states, even if:
- the reactants do not react under standard conditions
- a product is normally in a different state from its standard state, e.g. $H_2O(g)$ in combustion.

Make sure the equation shows the formation of 1 mol of product here. It is $\frac{1}{2}O_2$ because ethanol molecules only contain one oxygen atom.

Now try this

The standard enthalpy change of formation of hydrogen iodide, HI(g), is +26.5 kJ mol^{-1}.
Write an equation, and draw an enthalpy level diagram, to represent this change. **(3 marks)**

Measuring enthalpy changes

You can determine enthalpy changes experimentally by measuring the amount of energy transferred between the reactants and the surroundings.

Defining a standard enthalpy change

Standard enthalpy change of neutralisation, $\Delta H^{\ominus}_{neut}$, is:

- the enthalpy change measured at 100 kPa and a specified temperature, usually 298 K
- when one mole of water is produced by the neutralisation of an acid with an alkali
- with all solutions at 1 mol dm^{-3}.

The ionic equation for the reaction between a strong acid and a strong alkali is:

$$H^+(aq) + OH^-(aq) \rightarrow H_2O(l)$$

Enthalpy changes in solution

This apparatus can be used to measure enthalpy changes involving solutions.

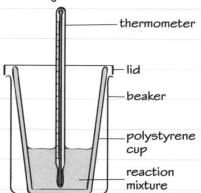

- thermometer
- lid
- beaker
- polystyrene cup
- reaction mixture

(a) Define the term standard enthalpy change of combustion, $\Delta H^{\ominus}_c$.   **(3 marks)**

The enthalpy change measured at 100 kPa and a specified temperature, usually 298 K, when one mole of a substance is completely burned in oxygen.

(b) A spirit burner containing ethanol was used to heat 100.0 cm^3 of water in a container by 20.1 K. If 0.46 g of ethanol (molar mass 46.0 g mol^{-1}) was burned, calculate the enthalpy change of combustion.   **(4 marks)**

Energy transferred to water $= 100.0 \times 4.18 \times 20.1$
$= 8402$ J

Energy change for fuel $= -8.40$ kJ

Amount of ethanol $= 0.46/46.0 = 0.010$ mol

$\Delta H_c = (-8.40)/0.010 = -840$ kJ mol^{-1}

The accepted value is -1367.3 kJ mol^{-1}. The lower experimental value is mainly due to heat losses and incomplete combustion of the fuel.

Calculating energy transferred

You can calculate the energy transferred to or from a substance using this $Q = mc\Delta T$

- Q = energy transferred in J
- m = mass of substance in g
- c = specific heat capacity in J g^{-1} K^{-1}
  ΔT = temperature change in K.

The **specific heat capacity** for water is 4.18 J g^{-1} K^{-1}. You also use this value for aqueous solutions such as acids and alkalis.

An enthalpy change calculation

For example, 1.00 mol dm^{-3} of hydrochloric acid neutralised by 1.20 mol dm^{-3} sodium hydroxide:
- volume of mixture $= 25.0 + 25.0$
  $= 50.0$ cm^3
- mean temperature before mixing $= 20.0\,°C$
- maximum temperature reached $= 26.9\,°C$

$Q = 50.0 \times 4.18 \times 6.9 = 1442$ J

mass of mixture, assuming $(26.9 - 20.0) = +6.9$ K
1 g cm^{-3} (the same as for water)

1442 J (1.442 kJ) of thermal energy was transferred to the water, so the energy change for the reactants was -1.442 kJ.

Enthalpy changes are measured in kJ mol^{-1}, so you need to know the amount of water formed:

amount of water = amount of acid neutralised
$= 1.00$ mol dm$^{-3} \times 0.0250$ dm$^3 = 0.0250$ mol
concentration of acid   25.0 cm^3/1000

$\Delta H_{neut} = (-1.442)/0.0250 = -57.7$ kJ mol^{-1}

When excess powdered zinc was added to 40.0 cm^3 of 1 mol dm^{-3} copper(II) sulfate, the temperature increased by 35.0 K. Calculate the enthalpy change for the reaction in kJ mol^{-1} of copper sulfate. Give your answer to 3 significant figures.
($c = 4.18$ J g^{-1} K^{-1})   **(4 marks)**

Ignore the mass of zinc and assume that 1 cm^3 of solution has a mass of 1 g.

Enthalpy cycles

Enthalpy changes may be determined indirectly using Hess's Law.

Hess's Law

Hess's Law states that the enthalpy change for a reaction is independent of the pathway taken.

It applies to reactions in which the initial and final conditions are the same for each pathway.

Hess's Law lets you calculate enthalpy changes for reactions that are difficult to achieve.

Elements in their standard states

Standard enthalpy change of formation, $\Delta H_f^{\ominus}$, is:

☑   the enthalpy change

☑   measured at 100 kPa and a specified temperature, usually 298 K

☑   when one mole of a substance is formed from its elements in their standard states.

$\Delta H_f^{\ominus}$ for an element in its standard state is zero.

Enthalpy cycles from formation data

Ethene reacts with steam to form ethanol, but not under standard conditions. To construct an enthalpy cycle to find $\Delta H_r^{\ominus}$ for a reaction, first write its equation:

$$C_2H_4(g) + H_2O(l) \rightarrow CH_3CH_2OH(l)$$

Underneath, write the elements in the amounts needed to form the substances in the equation:

$$2C(s) + 3H_2(g) + \tfrac{1}{2}O_2(g)$$

Draw and label arrows from the elements to the substances on both sides:

$$C_2H_4(g) + H_2O(l) \xrightarrow{\Delta H_r^{\ominus}} CH_3CH_2OH(l)$$

with arrows labelled A and B to

$$2C(s) + 3H_2(g) + \tfrac{1}{2}O_2(g)$$

Enthalpy change A = $\Delta H_f^{\ominus}[C_2H_4(g)] + \Delta H_f^{\ominus}[H_2O(l)]$
Enthalpy change B = $\Delta H_f^{\ominus}[CH_3CH_2OH(l)]$

Using Hess's Law: $\Delta H_r^{\ominus} = -A + (+B)$

   going against the arrow   going with the arrow

Enthalpy cycles from combustion data

Ethene is formed from carbon and hydrogen, but it would be difficult to achieve this synthesis directly. To construct an enthalpy cycle to find $\Delta H_r^{\ominus}$ for this reaction, first write its equation:

$$2C(s) + 2H_2(g) \rightarrow C_2H_4(g)$$

These substances react with oxygen, so write the combustion products underneath the equation:

$$2CO_2(g) + 2H_2O(l)$$

Draw and label arrows from both sides to the combustion products, adding the $3O_2$ needed:

$$3O_2(g) + 2C(s) + 2H_2(g) \xrightarrow{\Delta H_f^{\ominus}} C_2H_4(g) + 3O_2(g)$$

with arrows labelled A and B to

$$2CO_2(g) + 2H_2O(l)$$

Enthalpy change A = $2\Delta H_c^{\ominus}[C(s)] + 2\Delta H_c^{\ominus}[H_2(g)]$
Enthalpy change B = $\Delta H_c^{\ominus}[C_2H_4(g)]$

Using Hess's Law: $\Delta H_r^{\ominus} = +A + (-B)$

   going with the arrow   going against the arrow

Worked example

When heated, calcium carbonate decomposes to form calcium oxide and carbon dioxide. The enthalpy change for this reaction is difficult to determine directly. However, calcium carbonate and calcium oxide both react with hydrochloric acid. Draw an enthalpy cycle that could be used to determine $\Delta H_r^{\ominus}$ for this reaction. **(1 mark)**

$$2HCl(aq) + CaCO_3(s) \xrightarrow{\Delta H_r^{\ominus}} CaO(s) + CO_2(g) + 2HCl(aq)$$

with arrows labelled A and B to

$$CaCl_2(aq) + CO_2(g) + H_2O(l)$$

The reaction is thermal decomposition:

$$CaCO_3(s) \rightarrow CaO(s) + CO_2(g)$$

It is endothermic and you need to heat the calcium carbonate continually for the reaction to proceed. This makes it difficult to determine the enthalpy change of reaction directly.

Now try this

Draw enthalpy cycles that could be used to determine $\Delta H_r^{\ominus}$ for:
(a) The enthalpy change of reaction for: $NH_3(g) + HCl(g) \rightarrow NH_4Cl(s)$. **(1 mark)**
(b) The enthalpy change of formation of methane, CH_4. **(1 mark)**

Using enthalpy cycles

You can use enthalpy cycles with given data to calculate enthalpy changes of reaction.

Values for standard enthalpy changes

You need to be sure about the chemical change that leads to a quoted standard enthalpy change. Look at this reaction, for example:

$$C(s) + O_2(g) \rightarrow CO_2(g)$$

You could describe the standard enthalpy change as either $\Delta H_f^\ominus[CO_2(g)]$ or $\Delta H_c^\ominus[C(s)]$. The value of both is $-393.5\,\text{kJ mol}^{-1}$.

Calculations using formation data

The previous page shows how to construct the enthalpy cycle for the reaction between ethene and steam:

$$C_2H_4(g) + H_2O(l) \xrightarrow{\Delta H_r^\ominus} CH_3CH_2OH(l)$$

with A and B arrows to:

$$2C(s) + 3H_2(g) + \tfrac{1}{2}O_2(g)$$

Enthalpy change A = $\Delta H_f^\ominus[C_2H_4(g)] + \Delta H_f^\ominus[H_2O(l)]$
Enthalpy change B = $\Delta H_f^\ominus[CH_3CH_2OH(l)]$
Using supplied data:
A = $+52.2 + (-285.8) = -233.6\,\text{kJ mol}^{-1}$
B = $-277.1\,\text{kJ mol}^{-1}$
Using Hess's Law: $\Delta H_r^\ominus = -A + B$
$= -(-233.6) + (-277.1) = 233.6 - 277.1$
$= -43.5\,\text{kJ mol}^{-1}$

If the data given are all standard enthalpy changes of formation, you can calculate the standard enthalpy change of reaction like this:
$\Delta H_r^\ominus = (\Sigma\Delta H_f^\ominus[\text{products}]) - (\Sigma\Delta H_f^\ominus[\text{reactants}])$

means 'sum of' (add them all up)

Calculations using combustion data

The previous page shows how to construct the enthalpy cycle for the formation of ethene from carbon and hydrogen:

$$3O_2(g) + 2C(s) + 2H_2(g) \xrightarrow{\Delta H_f^\ominus} C_2H_4(g) + 3O_2(g)$$

with A and B arrows to:

$$2CO_2(g) + 2H_2O(l)$$

Enthalpy change A = $2\Delta H_c^\ominus[C(s)] + 2\Delta H_c^\ominus[H_2(g)]$
Enthalpy change B = $\Delta H_c^\ominus[C_2H_4(g)]$
Using supplied data:
A = $2(-393.5) + 2(-285.8) = -1358.6\,\text{kJ mol}^{-1}$
B = $-1410.8\,\text{kJ mol}^{-1}$
Using Hess's Law: $\Delta H_f^\ominus = +A - B$
$= -1358.6 - (-1410.8) = -1358.6 + 1410.8$
$= +52.2\,\text{kJ mol}^{-1}$

If the data given are all standard enthalpy changes of combustion, you can calculate the standard enthalpy change of reaction like this:
$\Delta H_r^\ominus = (\Sigma\Delta H_c^\ominus[\text{reactants}]) - (\Sigma\Delta H_c^\ominus[\text{products}])$

opposite to the equation on the left

Worked example

Calculate the standard enthalpy change of reaction for the thermal decomposition of calcium carbonate:
$CaCO_3(s) \rightarrow CaO(s) + CO_2(g)$
Use the data in the table below. **(2 marks)**

Substance	$\Delta H_f^\ominus/\text{kJ mol}^{-1}$
$CaCO_3(s)$	-1206.9
$CaO(s)$	-635.1
$CO_2(g)$	-393.5

$\Delta H_r^\ominus = (\Sigma\Delta H_f^\ominus[\text{products}]) - (\Sigma\Delta H_f^\ominus[\text{reactants}])$
$= -635.1 + (-393.5) - (-1206.9)$
$= -635.1 - 393.5 + 1206.9$
$= +178.3\,\text{kJ mol}^{-1}$

$$CaCO_3(s) \xrightarrow{+178.3} CaO(s) + CO_2(g)$$

-1206.9 and $-635.1 - 393.5 = -1028.6$ arrows to:

$$Ca(s) + C(s) + 1\tfrac{1}{2}O_2(g)$$

Now try this

Calculate the standard enthalpy change of reaction for the complete combustion of ethanol:
$CH_3CH_2OH(l) + 3O_2(g) \rightarrow 2CO_2(g) + 3H_2O(l)$
(2 marks)

You can find all the enthalpy of formation data you need on this page.

Mean bond enthalpy calculations

Mean bond enthalpies can be used to calculate enthalpy changes of reaction.

Bond enthalpy

Bond enthalpy, ΔH_B, is:

- the enthalpy change
- when one mole of a bond in the gaseous state is broken.

For a diatomic molecule, XY, it is the enthalpy change for this reaction:

$$XY(g) \rightarrow X(g) + Y(g)$$

For example: bond enthalpies are endothermic

$$HCl(g) \rightarrow H(g) + Cl(g) \quad \Delta H_B = +431\,kJ\,mol^{-1}$$

$$H_2(g) \rightarrow 2H(g) \qquad\qquad \Delta H_B = +436\,kJ\,mol^{-1}$$

$$Cl_2(g) \rightarrow 2Cl(g) \qquad\qquad \Delta H_B = +242\,kJ\,mol^{-1}$$

homolytic bond fission

Mean bond enthalpy

Bond enthalpy varies with:

- the particular bond involved
- the chemical environment of the bond.

For example, there are four different C–H bond enthalpies in methane, CH_4:

- $CH_4(g) \rightarrow CH_3(g) + H(g) \quad \Delta H_B = +423\,kJ\,mol^{-1}$
- $CH_3(g) \rightarrow CH_2(g) + H(g) \quad \Delta H_B = +480\,kJ\,mol^{-1}$
- $CH_2(g) \rightarrow CH(g) + H(g) \quad \Delta H_B = +425\,kJ\,mol^{-1}$
- $CH(g) \rightarrow C(g) + H(g) \qquad \Delta H_B = +335\,kJ\,mol^{-1}$

The mean of these enthalpies is $415.75\,kJ\,mol^{-1}$, corresponding to: $\frac{1}{4}CH_4(g) \rightarrow \frac{1}{4}C(g) + H(g)$

Mean bond enthalpy is the enthalpy change when one mole of a bond in the gaseous state, averaged out over many different molecules, is broken.

Using mean bond enthalpies

To calculate an estimate of $\Delta H_r^\ominus$:

 Add up the mean bond enthalpies for all the bonds broken in the reactants, Σ(bonds broken).

 Add up the mean bond enthalpies for all the bonds made in the products, Σ(bonds made).

 $\Delta H_r^\ominus = \Sigma$(bonds broken) $- \Sigma$(bonds made)

A simple example

What is the enthalpy change of reaction for this reaction? $H_2(g) + Cl_2(g) \rightarrow 2HCl(g)$

Using the mean bond enthalpy values above:

 Σ(bonds broken) $= 436 + 242$
$$= 678\,kJ\,mol^{-1}$$

 Σ(bonds made) $= 2 \times 431$
$$= 862\,kJ\,mol^{-1}$$

 $\Delta H_r^\ominus = 678 - 862 = -184\,kJ\,mol^{-1}$

Worked example

Calculate the mean bond enthalpy for the C–H bond, to 3 significant figures, using these data:

$$CH_4(g) + 2H_2O(l) \rightarrow 4H_2(g) + CO_2(g)$$

$\Delta H_r^\ominus = +253\,kJ\,mol^{-1}$ **(3 marks)**

Bond	Mean bond enthalpy/kJ mol⁻¹
O–H	463
H–H	436
C=O	743

Σ(bonds broken) $= 4E(C–H) + 4E(O–H)$
$$= 4E(C–H) + 1852\,kJ\,mol^{-1}$$

Σ(bonds made) $= 4E(H–H) + 2E(C=O)$
$$= (4 \times 436) + (2 \times 743)$$
$$= 1744 + 1486$$
$$= 3230\,kJ\,mol^{-1}$$

$\Delta H_r^\ominus = \Sigma$(bonds broken) $- \Sigma$(bonds made)

$+253 = 4E(C–H) + 1852 - 3230$

$4E(C–H) = +253 - 1852 + 3230 = 1631$

Mean bond enthalpy $= \dfrac{1631}{4} = 408\,kJ\,mol^{-1}$

Differences in values

When you use mean bond enthalpies you may obtain an answer different from the one given in a Data Book. This can be because:

- The bond enthalpies in the substances involved may be different from the means.
- Bond enthalpies assume the gaseous state but the substances may be solid or liquid.

Now try this

1. The mean bond enthalpy for the C–H bond is given in a Data Book as $435\,kJ\,mol^{-1}$. Give two reasons why this differs from the value calculated in the Worked Example.   **(2 marks)**

2. Calculate the enthalpy change for the combustion of hydrogen: $2H_2(g) + O_2(g) \rightarrow 2H_2O(l)$
Use $498\,kJ\,mol^{-1}$ for the O=O bond, and mean enthalpies in the Worked Example.   **(3 marks)**

Changing reaction rate

Rate of reaction is determined by changes in concentration of a reactant or a product per unit time.

Measuring rate of reaction

In general:

$$\text{rate of reaction} = \frac{\text{change in concentration}}{\text{time taken for the change}}$$

You can measure the **rate** of a reaction by:
- measuring how fast a reactant is used up, or
- measuring how fast a product is formed.

As the reaction proceeds, the concentration of a product increases (below) and the concentration of a reactant decreases (right).

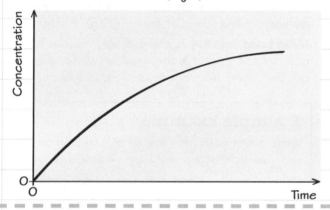

Rate from the gradient of a graph

When you plot a graph of concentration of a substance against time, you can measure its **gradient** at a particular time in the reaction:

☑ Draw a tangent to the curve at that time.

☑ Complete a triangle (sides of at least 8 cm will do).

☑ calculate: gradient = $\frac{y}{x}$

The greater the gradient, the greater the rate of reaction.

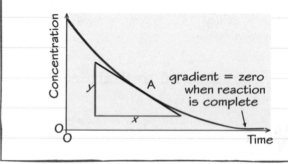

gradient = zero when reaction is complete

Collision theory

For two particles to react, they must:
- collide with each other, and
- the collision must have sufficient energy.

The **activation energy**, E_a, is the minimum amount of energy that colliding particles need for a reaction to happen. It is involved in breaking bonds in the reactants.

For many compounds, particles must also collide in the correct **orientation**. For example:

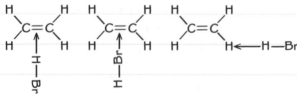

possible reaction   no reaction   no reaction

Rate and surface area of solid reactants

A fine powder has a greater **surface area** than the same mass of the substance in big lumps.

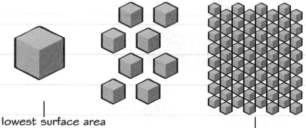

lowest surface area

greatest surface area

The greater the surface area:
- the greater the rate of collision between reactant particles
- the greater the rate of reaction.

Worked example

Calcium carbonate reacts with hydrochloric acid. Explain why changing the concentration of the acid may affect the rate of reaction. **(2 marks)**

If the concentration of the acid is increased, there are more acid particles in the same volume. This increases the rate of collisions between reactant particles, increasing the rate.

Now try this

Explain, in terms of reactant particles, why the rate of reaction decreases when the concentration of a solution is decreased, and when a lump is used instead of a powder. **(4 marks)**

The **frequency** of successful collisions (the ones with sufficient energy to react) increases, but not the **proportion** of collisions that are successful.

Maxwell–Boltzmann model

The Maxwell–Boltzmann model describes the distribution of molecular energies in gases.

The Maxwell–Boltzmann distribution

The distribution of molecular energies in a sample of gas can be represented by a graph like this:

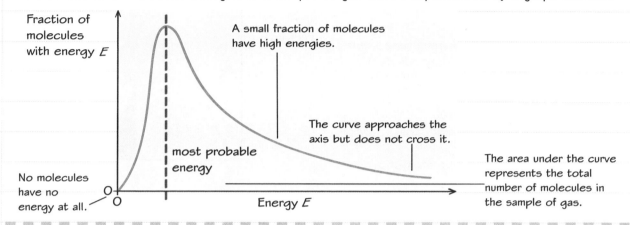

Fraction of molecules with energy E

A small fraction of molecules have high energies.

The curve approaches the axis but does not cross it.

most probable energy

The area under the curve represents the total number of molecules in the sample of gas.

No molecules have no energy at all.

O

Energy E

Curves at different temperatures

The distribution of molecular energies changes as the temperature increases from T_1 to T_2.

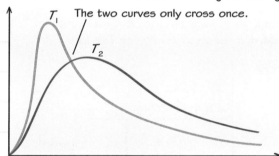

T_1 The two curves only cross once.

T_2

At higher temperatures, the curve:

- is displaced to the right
- has a higher most probable energy
- is lower than before
- has the same area as before.

Worked example

(a) The graph shows the distribution of molecular energies in a sample of gas. Draw a line to represent the distribution at a higher temperature. **(2 marks)**

Second line is lower, displaced to the right, and only crosses the existing line once.

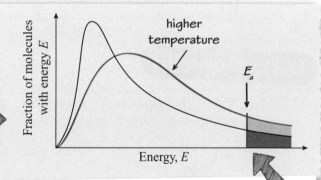

higher temperature

E_a

Fraction of molecules with energy E

Energy, E

(b) Use the graphs to help you explain why the rate of reaction between gaseous particles increases with temperature. **(3 marks)**

At higher temperatures, the fraction of molecules with the activation energy or more increases. This increases the frequency of successful collisions, so the rate of reaction increases.

The student has added an activation energy, E_a, and shaded the areas under the curves to show the number of molecules with sufficient energy to react.

Note that most particles have less energy than E_a, so their collisions will be unsuccessful.

Now try this

On the same axes, draw graphs to show the distribution of molecular energies in a sample of gas at two temperatures. Label the axes and indicate the fraction of molecules with sufficient energy to react. **(5 marks)**

Catalysts

Catalysts affect the rate of a reaction without being used up in the reaction.

Lowering activation energy

A **catalyst** provides an alternative reaction route with a lower activation energy than the original uncatalysed route.

This increases the fraction of gaseous molecules with sufficient energy to react if they collide.

Alternative pathways

Catalysts are described as either:
- heterogeneous – reactants and catalyst are in different phases, or
- homogeneous – reactants and catalyst are in the same phase.

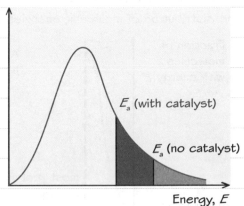

Reaction profiles

A **reaction profile** is similar to an enthalpy level diagram but with these extra features:

- horizontal axis labelled
- the activation energy, E_a, is shown.

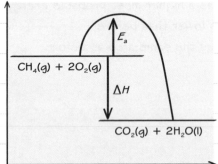

Enthalpy level diagrams

An **enthalpy level diagram** shows the changes in enthalpy during a reaction (see page 65 for more detail about this).

This enthalpy diagram corresponds to the exothermic reaction profile on the left.

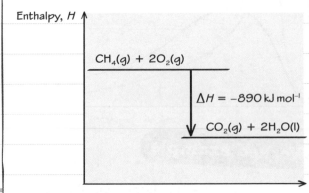

Worked example

This reaction profile represents an uncatalysed reaction. On the diagram, draw the profile for the same reaction in the presence of a catalyst, and indicate the activation energy. **(2 marks)**

The enthalpies of the reactants and products stay the same, but E_a for the catalysed reaction is lower. This means that, at a given temperature, a greater proportion of collisions between reactant particles will be successful.

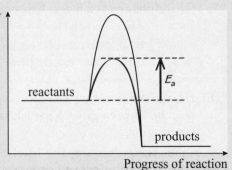

Now try this

1 Draw a Maxwell–Boltzmann distribution curve and show how the fraction of molecules with sufficient energy to react changes in the presence of a catalyst. **(5 marks)**
2 Draw a reaction profile for an endothermic reaction in the presence and absence of a catalyst. **(3 marks)**

Similar to an exothermic reaction profile but with products higher than reactants.

Dynamic equilibrium 1

Many reactions are reversible, and they can reach a state of dynamic equilibrium.

Reversible reactions

Reactions in which ΔH is large and negative go to completion – they cannot be easily reversed.

Reversible reactions:

- do not go to completion
- contain detectable amounts of reactants mixed with the products
- are shown in equations by $\rightleftharpoons$ instead of $\rightarrow$

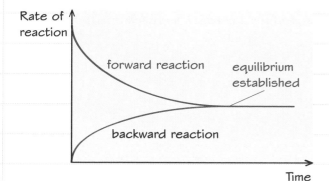

Establishing equilibrium

For example, the reaction between hydrogen and iodine to form hydrogen iodide is reversible:

$$H_2(g) + I_2(g) \rightleftharpoons 2HI(g)$$

In a sealed container of hydrogen and iodine vapour, the **forward reaction** begins:

$$H_2(g) + I_2(g) \rightarrow 2HI(g)$$

As this reaction proceeds:

- The concentrations of $H_2(g)$ and $I_2(g)$ decrease.
- The rate of the forward reaction decreases.
- The concentration of $HI(g)$ increases.

As this happens, the rate of the **backward reaction** increases:

$$2HI(g) \rightarrow H_2(g) + I_2(g)$$

The rates of the two reactions eventually become the same, and **equilibrium** is established (see graph on left).

Effect of changing concentration

The **position of equilibrium** changes if you change the concentration of one of the reacting substances in a system at equilibrium:

Concentration of:		Position of equilibrium
Reactant	Product	
increased	decreased	moves to the right
decreased	increased	moves to the left

This mixture reaches equilibrium in a beaker:

$$2CrO_4{}^{2-}(aq) + 2H^+(aq) \rightleftharpoons Cr_2O_7{}^{2-}(aq) + H_2O(l)$$
    yellow   orange

Change in $H^+(aq)$	Position of equilibrium	Observation
↑	→	turns orange
↓	←	turns yellow

     alkali added

Dynamic equilibrium

In a **dynamic equilibrium**:

☑ The forward and backward reactions carry on continuously – the 'dynamic' bit of the name.

☑ The rate of the forward reaction equals the rate of the backward reaction.

☑ The concentrations of reactants and products remain constant.
     These do not have to be the same as each other!

Remember, for dynamic equilibrium to be established:

☑ The reaction must be reversible, and

☑ The reaction mixture must be in a closed container, so that no substances enter or leave.

Worked example

Nitrogen reacts with hydrogen to form ammonia:
$$N_2(g) + 3H_2(g) \rightleftharpoons 2NH_3(g)$$
Predict the effect of increasing the pressure of this mixture of gases at equilibrium. **(2 marks)**

The position of equilibrium will move to the right because there are fewer moles of gas on the right hand side of the equation. The equilibrium concentration of ammonia will increase.

Now try this

This is one of the reactions that can happen in a catalytic converter:

$$2NO(g) + 2CO(g) \rightleftharpoons N_2(g) + 2CO_2(g)$$

Predict the effect on the position of equilibrium if the pressure is reduced. **(2 marks)**

Pressure changes have no effect on the position of equilibrium in this reaction: $H_2(g) + I_2(g) \rightleftharpoons 2HI(g)$

Dynamic equilibrium 2

The equilibrium constant for a reaction at equilibrium remains the same unless the temperature is changed.

Temperature and position of equilibrium

If you increase the temperature of a reaction mixture in equilibrium:
- The rates of the forward reaction and backward reaction both increase, but
- the rate of the endothermic reaction increases by a greater ratio.

This means that an increase in temperature will move the position of equilibrium in the direction of the endothermic reaction. For example:

$$N_2(g) + 3H_2(g) \underset{\Delta H = +92\,kJ\,mol^{-1}}{\overset{\Delta H = -92\,kJ\,mol^{-1}}{\rightleftharpoons}} 2NH_3(g)$$

endothermic change

Increasing the temperature will move the position of equilibrium to the left.

Catalysts and equilibria

If you add a catalyst to a reaction mixture that is in equilibrium:
- The rates of the forward reaction and backward reaction both increase, and
- by the same ratio.

  If the forward rate doubles, so does the backward rate.

Adding a catalyst does not change the position of equilibrium, but it does reduce the time needed for a reaction to reach equilibrium.

K_c for a homogeneous system

The **equilibrium constant** expressed in terms of concentrations has the symbol K_c.

For a mixture of substances A, B, C and D:

$$aA + bB \rightleftharpoons cC + dD$$

$$K_c = \frac{[C]^c[D]^d}{[A]^a[B]^b}$$

concentration of A in $mol\,dm^{-3}$

This expression defines the **equilibrium law**.

Two different systems

In a **homogeneous system**:

☑ All the components are in the same phase or state.

For example: $H_2(g) + I_2(g) \rightleftharpoons 2HI(g)$

In a **heterogeneous system**:

☑ Two or more components are in different phases or states.

For example: $CaCO_3(s) \rightleftharpoons CaO(s) + CO_2(g)$

Worked example

(a) Nitrogen reacts with hydrogen to form ammonia: $N_2(g) + 3H_2(g) \rightleftharpoons 2NH_3(g)$
Write an expression for the equilibrium constant, K_c. **(1 mark)**

$$K_c = \frac{[NH_3(g)]^2}{[N_2(g)][H_2(g)]^3}$$

(b) Calcium carbonate decomposes when heated: $CaCO_3(s) \rightleftharpoons CaO(s) + CO_2(g)$

Write an expression for the equilibrium constant, K_c. **(1 mark)**

$$K_c = [CO_2(g)]$$

This is a homogeneous system, so all its components appear in the expression for K_c.

Going from the chemical equation to K_c:

$$N_2(g) + 3H_2(g) \rightleftharpoons 2NH_3(g)$$

$[N_2(g)]$ raised to the power 1 $[H_2(g)]$ raised to the power 3 $[NH_3(g)]$ raised to the power 2

This is a heterogeneous system. At a given temperature, the equilibrium concentrations of solids and liquids are constant. Their concentrations do not appear in the expression for K_c, only the concentrations of gases or solutions raised to appropriate powers.

Now try this

Write expressions for the equilibrium constant K_c for the following reactions.
(a) $H_2(g) + I_2(g) \rightleftharpoons 2HI(g)$ **(1 mark)**
(b) $2NO_2(g) \rightleftharpoons N_2O_4(g)$ **(1 mark)**
(c) $H_2O(l) \rightleftharpoons H^+(aq) + OH^-(aq)$ **(1 mark)**
(d) $C(s) + H_2O(g) \rightleftharpoons H_2(g) + CO(g)$ **(1 mark)**

In (c), remember that an aqueous solution is not the same as a liquid.

Industrial processes

In many industrial processes a compromise between the yield and the rate of reaction is necessary.

Yield

The factors affecting the equilibrium **yield** of a product from a reversible reaction include the:

- ✓ temperature of the reaction mixture
- ✓ concentration of reactants
- ✓ pressure, if gases are present.

Although catalysts do not affect the position of equilibrium, they do reduce the time taken to reach equilibrium.

The reaction conditions chosen in an industrial process are not necessarily the ones that will produce the greatest equilibrium yield.

Rate of reaction

Factors affecting rate of reaction include:

- ✓ temperature of the reaction mixture
- ✓ concentration of reactants
- ✓ particle size of solid reactants
- ✓ presence of a suitable catalyst.

An increase in the pressure of reacting gases may increase the rate of reaction, but not if:

- the uncatalysed rate is very low, and
- a catalyst is present, and
- the catalyst's active sites are all occupied.

The Haber process

The Haber process manufactures ammonia:

$$N_2(g) + 3H_2(g) \rightleftharpoons 2NH_3(g) \quad \Delta H = -92\,kJ\,mol^{-1}$$

The forward reaction is exothermic.

A high equilibrium yield is favoured by:

- a low temperature
- a high pressure. ——— fewer moles of gas on the right

Iron is used as a catalyst because it is cheaper than other choices such as platinum or tungsten.

The pressure chosen is high enough to obtain a reasonable yield, but not so high that it is expensive to achieve.

The temperature chosen is high enough to achieve a reasonable rate of reaction, but not so high that it would reduce the equilibrium yield.

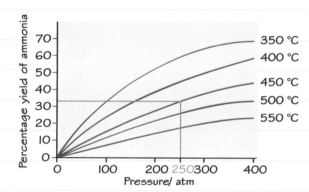

The conditions chosen are usually:

- 250 atm (25 MPa) pressure
- 450 °C temperature.

To reduce the time needed, the reaction is not allowed to reach equilibrium and the yield is 15%.

Worked example

Methanol can be manufactured from carbon monoxide and hydrogen:

$$CO(g) + 2H_2(g) \rightleftharpoons CH_3OH(g) \quad \Delta H = -91\,kJ\,mol^{-1}$$

(a) Explain why a low temperature would favour a high equilibrium yield of methanol. **(2 marks)**

The forward reaction is exothermic, so the position of equilibrium moves to the right if the temperature is reduced.

The temperature must not be too low or the rate will not be great enough to be economic.

(b) Explain why a high pressure would favour a high equilibrium yield of methanol. **(2 marks)**

There are fewer moles of gas on the right, so the position of equilibrium moves to the right if the pressure is increased.

(c) State two reasons why a very high pressure would be uneconomic. **(2 marks)**

A strong pressure vessel would be needed and a lot of energy would be needed to pressurise it.

Now try this

One stage in the manufacture of sulfuric acid involves: $SO_2(g) + \frac{1}{2}O_2(g) \rightleftharpoons SO_3(g) \quad \Delta H = -96\,kJ\,mol^{-1}$

(a) Make qualitative predictions of the temperature and pressure that would favour a high equilibrium yield of sulfur trioxide, SO_3. **(4 marks)**

(b) Suggest why a temperature of around 450 °C is used in the process. **(2 marks)**

Exam skills 7

This exam-style question uses knowledge and skills you have already revised. Have a look at pages 73–75 for a reminder about **dynamic equilibria** and **industrial processes**.

Worked example

Hydrogen is useful as a fuel and for making ammonia by the Haber process. It can be made in different ways.

(a) Carbon, in the form of coke, reacts with steam:

$$C(s) + 2H_2O(g) \rightleftharpoons CO_2(g) + 2H_2(g)$$

Write an expression for the equilibrium constant, K_c, for this reaction. **(1 mark)**

$$K_c = \frac{[CO_2(g)][H_2(g)]^2}{[H_2O(g)]^2}$$

(b) Hydrogen can also be manufactured from methane and steam:

$$CH_4(g) + H_2O(g) \rightleftharpoons CO(g) + 3H_2(g)$$
$$\Delta H^{\ominus} = +205\,kJ\,mol^{-1}$$

(i) Explain the effect of increasing the temperature on the equilibrium yield of hydrogen. **(2 marks)**

The equilibrium yield increases as the temperature increases because the forward reaction is endothermic.

(ii) Explain the effect of increasing the pressure on the equilibrium yield of hydrogen. **(2 marks)**

The equilibrium yield decreases as the pressure increases because there are fewer molecules of gas on the left hand side of the equation.

(iii) Explain why excess steam is used in the process, rather than excess methane. **(2 marks)**

An excess of reactants increases the rate of the forward reaction, moving the position of equilibrium to the right. Methane is more expensive, so it is better to use an excess of steam.

(d) Other than cost alone, explain one advantage of using catalysts in industrial chemical processes. **(2 marks)**

Catalysts allow reactions to happen at lower temperatures, which means that less fuel is needed to heat up the reactants.

(e) Explain why chemical reactions in industrial processes are usually not allowed to reach equilibrium. **(2 marks)**

There is a balance to strike between reaction rate and yield. The more product that can be made in a given amount of time, the more economically viable the process is likely to be.

Command word: Write

The command word **write** is used if a question asks you to provide an equation.

Maths skills Carbon is present in the solid state, so it is not included in the expression for K_c.
Remember that square brackets mean concentration in $mol\,dm^{-3}$, so do not omit them or write round brackets instead.

The value quoted for $\Delta H^{\ominus}$ refers to the forward reaction, unless stated otherwise.

It is implied in the **explain** command word that you also need to state what the effect on the equilibrium yield is.

You could also answer in terms of maintaining a constant value for K_c at a given temperature. For this to happen, the concentrations of products must decrease and the concentrations of reactants must increase.

You could also mention that an excess of steam ensures that a greater proportion of the methane reacts, or that steam is renewable but methane is not.

You could also mention that catalysts provide alternative reaction routes with lower activation energies.
These may have higher atom economies than the uncatalysed processes, so fewer raw materials are needed.

You could give other reasons instead, such as:
• The products are removed as they form, so the system is not closed.
• Unreacted substances may be recycled to an earlier part of the process.

An explanation involving cost alone, with no supporting argument, is not enough.

Partial pressures and K_p

Equilibrium constants for reactions involving gases can be expressed in terms of partial pressures.

Partial pressures

In a mixture of gases, the **partial pressure** of a gas is the pressure that the gas would exert on its own under the same conditions.

For example, the standard pressure of air is 100 kPa. The table shows the approximate partial pressures of the main components of air.

Gas	% in air	Mole fraction	Partial pressure/kPa
N_2	78	0.78	78
O_2	21	0.21	21
Ar	0.9	0.009	0.9
CO_2	0.04	0.0004	0.04

Calculating a partial pressure

You can use this expression to calculate the partial pressure of a gas A:

$$p_A = x_A p_T$$

partial pressure of gas A   mole fraction of gas A   total pressure

You can use this expression to calculate the **mole fraction** of a gas A:

$$x_A = \frac{\text{amount of A}}{\text{total amount of gas}}$$

Mole fractions have no units, as the amounts of each gas are measured in the same units (mol).

Worked example

Hydrogen iodide, hydrogen and iodine may reach equilibrium: $2HI(g) \rightleftharpoons H_2(g) + I_2(g)$
4.0 mol of hydrogen iodide was sealed in a flask at 2.0 atm pressure and heated. At equilibrium, there was 3.6 mol of hydrogen iodide.

(a) Calculate the equilibrium amount of H_2.

Amount of HI reacted = 4.0 − 3.6 = 0.4 mol
Amount of H_2 produced = 0.4 ÷ 2 = 0.2 mol

(b) Calculate the partial pressure of hydrogen.

Amount of I_2 = amount of H_2 = 0.2 mol
Total amount of gas = 3.6 + 0.2 + 0.2
                  = 4.0 mol
Mole fraction of H_2 = 0.2 ÷ 4.0 = 0.05
Partial pressure of H_2 = 0.05 × 2.0 = 0.1 atm

The atmosphere, atm, is a unit of pressure:
1 atm = 101 325 Pa

From the balanced equation, you should see that 2 mol of HI produce 1 mol of H_2 (and 1 mol of I_2). This is why the amount of HI reacted is divided by 2 in the second line of the calculation.

Remember that mole fraction has no units, but the partial pressure of a gas has the same units as the total gas pressure.
The total pressure is equal to the sum of all the individual partial pressures.

Worked example

(a) Nitrogen reacts with hydrogen to form ammonia:
$N_2(g) + 3H_2(g) \rightleftharpoons 2NH_3(g)$
Write an expression for the equilibrium constant, K_p. **(1 mark)**

$$K_p = \frac{(p_{NH_3})^2}{(p_{N_2})(p_{H_2})^3}$$

(b) Calcium carbonate decomposes when heated:
$CaCO_3(s) \rightleftharpoons CaO(s) + CO_2(g)$
Write an expression for the equilibrium constant, K_p. **(1 mark)**

$$K_p = p_{CO_2}$$

Solids, liquids and solutions do not appear in an expression for K_p, only gases do.

Now try this

Write expressions for the equilibrium constant, K_p, for the following equilibria.

(a) $PCl_5(g) \rightleftharpoons PCl_3(g) + Cl_2(g)$ **(1 mark)**

(b) $2SO_2(g) + O_2(g) \rightleftharpoons 2SO_3(g)$ **(1 mark)**

(c) $NH_4Cl(s) \rightleftharpoons NH_3(g) + HCl(g)$ **(1 mark)**

Calculating K_c and K_p values

If equilibrium concentrations or partial pressures are known, the value of K_c or K_p can be calculated.

Units for K_c

To work out the units for K_c:

1. Write the expression for K_c.

2. Total the powers for the top concentrations.

3. Total the powers for the bottom concentrations.

4. Work out (answer 2) – (answer 3).

5. Apply answer 4 to the concentration units, $mol\,dm^{-3}$.

For example, $N_2(g) + 3H_2(g) \rightleftharpoons 2NH_3(g)$

$$K_c = \frac{[NH_3(g)]^2}{[N_2(g)][H_2(g)]^3}$$

total power is 2

total power is $(1 + 3) = 4$

$2 - 4 = -2$, so the units for K_c here are $mol^{-2}\,dm^6$.

$(-2) \times (+1) = -2$ $(-2) \times (-3) = 6$

Units for K_p

To work out the units for K_p:

1. Write the expression for K_p.

2. Total the powers for the top partial pressures.

3. Total the powers for the bottom partial pressures.

4. Work out (answer 2) – (answer 3).

5. Apply answer 4 to the pressure units, atm.

For example, $N_2(g) + 3H_2(g) \rightleftharpoons 2NH_3(g)$

$$K_p = \frac{(p_{NH_3})^2}{(p_{N_2})(p_{H_2})^3}$$

total power is 2

total power is $(1 + 3) = 4$

$2 - 4 = -2$, so the units for K_p here are atm^{-2}.

If there are equal numbers of particles on both sides of the equation, the value of K_c or K_p will have no units, so do point this out in an answer.

Worked example

(a) N_2O_4 and NO_2 may reach equilibrium:
$$N_2O_4(g) \rightleftharpoons 2NO_2(g)$$
1.00 mol of N_2O_4 was sealed in a $0.50\,dm^3$ flask. At equilibrium with NO_2, there was 0.60 mol of N_2O_4. Calculate the equilibrium concentrations of N_2O_4 and NO_2. **(2 marks)**

$[N_2O_4] = 0.60 \div 0.50 = 1.2\,mol\,dm^{-3}$
Amount of N_2O_4 reacted $= 1.00 - 0.60$
$\qquad\qquad\qquad\qquad = 0.40\,mol$
Amount of NO_2 produced $= 0.40 \times 2$
$\qquad\qquad\qquad\qquad = 0.80\,mol$
$[NO_2] = 0.80 \div 0.50 = 1.6\,mol\,dm^{-3}$

(b) $K_c = \dfrac{[NO_2(g)]^2}{[N_2O_4(g)]}$

Calculate the value for K_c, including its units. **(2 marks)**

$K_c = \dfrac{(1.6)^2}{1.2} = \dfrac{2.56}{1.2}$

$K_c = 2.1\,mol\,dm^{-3}$

Volume K_c and K_p

You can ignore the volume for a K_c calculation if the value of K_c has no units.

You can ignore the volume for a K_p calculation and just use the partial pressures at equilibrium.

The first Worked Example on page 77 shows how to work out partial pressures at equilibrium.

Take care to substitute the correct equilibrium concentrations (or partial pressures for K_p) into the expression.

Remember to raise each number to the correct power, as it can be easy to forget to square a value, for example.

Now try this

Sulfur dioxide reacts with oxygen to produce sulfur trioxide: $2SO_2(g) + O_2(g) \rightleftharpoons 2SO_3(g)$.
The equilibrium partial pressures are SO_2, 0.060 atm; O_2, 0.23 atm; SO_3, 1.7 atm.
(a) Write the expression for K_p. **(1 mark)**
(b) Calculate its value, giving the units, if any. **(3 marks)**

Changing K_c and K_p

The values of K_c or K_p for a reaction are only affected by changes in temperature, not by catalysts.

Exothermic changes

For a reversible reaction that is exothermic in the forward direction, K_c or K_p decreases with increasing temperature:

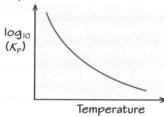

For example, $N_2(g) + 3H_2(g) \rightleftharpoons 2NH_3(g)$:

$$K_p = \frac{(p_{NH_3})^2}{(p_{N_2})(p_{H_2})^3} \quad \begin{array}{l}\text{—— decreases}\\[4pt]\text{—— increases}\end{array}$$

The position of equilibrium moves to the left.

Endothermic changes

For a reversible reaction that is endothermic in the forward direction, K_c or K_p increases with increasing temperature.

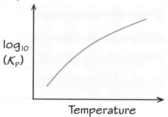

For example, $N_2O_4(g) \rightleftharpoons 2NO_2(g)$:

$$K_p = \frac{(p_{NO_2})^2}{(p_{N_2O_4})} \quad \begin{array}{l}\text{—— increases}\\[4pt]\text{—— decreases}\end{array}$$

The position of equilibrium moves to the right.

Effect of concentration

Changing the concentration of a reacting substance does not affect K_c. For example:

$Cl_2(aq) + H_2O(l) \rightleftharpoons HCl(aq) + HClO(aq)$:

$$K_c = \frac{[HCl(aq)_{eqm}][HClO(aq)_{eqm}]}{[Cl_2(aq)_{eqm}]}$$

$$Q_c = \frac{[HCl(aq)][HClO(aq)]}{[Cl_2(aq)]}$$

If $[Cl_2(aq)]$ is suddenly increased, Q_c decreases and is no longer equal to K_c.

$[Cl_2(aq)]$ will decrease, increasing $[HCl(aq)]$ and $[HClO(aq)]$, until $Q_c = K_c$ again.

Worked example

Hydrogen can be manufactured by reacting carbon with steam: $C(s) + H_2O(g) \rightleftharpoons CO(g) + H_2(g)$

$$K_p = \frac{(p_{co})(p_{H_2})}{(p_{H_2O})}$$

Explain why an increase in pressure leads to a decrease in the yield of hydrogen. **(2 marks)**

In terms of mole fractions and total pressure, p_T:

$$K_p = \frac{(x_{co})(x_{H_2})}{(x_{H_2O})} \times \frac{(p_T)^2}{p_T} = \frac{(x_{co})(x_{H_2})}{(x_{H_2O})} \times p_T$$

So, to keep K_p constant when the total pressure is increased, the mole fractions of CO and H_2 must decrease, and the mole fraction of H_2O must increase. This reduces the yield of H_2.

Reaction quotient, Q_c

The expression for the **reaction quotient** of a reaction is the same as the expression for K_c but the concentrations are not necessarily the ones at equilibrium.

Effect of adding a catalyst

Catalysts do not affect the position of equilibrium, or the value of K_c or K_p. Catalysts:

*   increase the rates of the forward and backward reactions by the same ratio
*   reduce the time taken to reach equilibrium
*   do not appear in expressions for K_c or K_p.

Remember: $p_A = x_A p_T$
where p_A is the partial pressure of A, x_A is the mole fraction of A, and p_T is the total pressure.

Now try this

Dinitrogen tetroxide exists in equilibrium with nitrogen dioxide: $N_2O_4(g) \rightleftharpoons 2NO_2(g)$

In terms of the expression for the equilibrium constant, K_p, and its value, and explain why an increase in pressure leads to a decrease in the yield of nitrogen dioxide. **(2 marks)**

Acids, bases and pH

The Brønsted–Lowry theory of acids and bases involves the transfer of hydrogen ions, $H^+(aq)$, which are also identified as protons.

Brønsted–Lowry acids

A **Brønsted–Lowry acid** is a **proton donor**.

For example:

$$HCl(aq) + H_2O(l) \rightleftharpoons H_3O^+(aq) + Cl^-(aq)$$

HCl acts as a Brønsted–Lowry acid because it donates a proton (hydrogen ion, H^+) to H_2O.

Acids may be classified according to the number of protons they can donate:

- Hydrochloric acid, HCl, is a **monobasic** acid.
- Sulfuric acid, H_2SO_4, is a **dibasic** acid.
- Phosphoric acid, H_3PO_4, is a **tribasic** acid.

You may also see monoprotic, diprotic and triprotic applied to acids in place of these terms.

Brønsted–Lowry bases

A **Brønsted–Lowry base** is a **proton acceptor**.

For example:

$$NH_3(aq) + H_2O(l) \rightleftharpoons NH_4^+(aq) + OH^-(aq)$$

NH_3 acts as a Brønsted–Lowry base because it accepts a proton (hydrogen ion, H^+) from H_2O.

Bases may be classified according to the number of protons they can accept:

- Ammonia, NH_3, is a monobasic base.
- The carbonate ion, CO_3^{2-}, is a dibasic base.
- The phosphate ion, PO_4^{3-}, is a tribasic base.

These ions are found in soluble compounds such as sodium carbonate and sodium phosphate.

Conjugate acid–base pairs

When an acid donates a proton, the species formed is the **conjugate base** of the acid.

When a base accepts a proton, the species formed is the **conjugate acid** of the base.

For example, nitric acid and ethanoic acid react together:

$$HNO_3 + CH_3COOH \rightleftharpoons NO_3^- + CH_3COOH_2^+$$

proton donor (acid)   proton acceptor (base)   conjugate base   conjugate acid

HNO_3 and NO_3^-, and CH_3COOH and $CH_3COOH_2^+$, are the two **conjugate acid–base pairs** here.

Amphoteric substances

An **amphoteric substance** can act both as an acid and as a base.

Water is an amphoteric substance. It can:

☑ donate protons, for example:

$$NH_3(aq) + H_2O(l) \rightleftharpoons NH_4^+(aq) + OH^-(aq)$$

☑ accept protons, for example:

$$HCl(aq) + H_2O(l) \rightleftharpoons H_3O^+(aq) + Cl^-(aq)$$

The **oxonium** ion

Worked example

(a) Define pH.   **(1 mark)**

$$pH = -\log_{10}[H^+]$$

(b) Calculate the pH of a solution that contains $0.01 \, mol \, dm^{-3}$ $H^+(aq)$.   **(1 mark)**

$$pH = -\log_{10}(0.01) = -(-2) = 2$$

(c) Calculate the hydrogen ion concentration of a solution at pH 4.   **(1 mark)**

$$[H^+(aq)] = 10^{-pH}$$

$$[H^+(aq)] = 10^{-4} = 1 \times 10^{-4} \, mol \, dm^{-3}$$

This could also be answered as $pH = -\log[H^+]$, with $[H_3O^+]$ in place of $[H^+]$, or even in words ('the negative logarithm of the hydrogen ion concentration'). $pH = -\log_e[H^+]$ and $pH = -\ln[H^+]$ are <u>incorrect</u>.

 Maths skills Use the **log** or **lg** button on your calculator.
Check that you get: $\log(10) = 1$

Maths skills Use the 10^x button on your calculator (or enter 10 y^x followed by the value of $-pH$).
Check that you get: $10^{-1} = 0.1$

Now try this

1 Consider this reaction:
$$H_2SO_4 + HNO_3 \rightleftharpoons HSO_4^- + H_2NO_3^+$$
(a) Explain which substance, H_2SO_4 or HNO_3, is acting as a Brønsted–Lowry acid.   **(1 mark)**

(b) Identify the two conjugate acid–base pairs.   **(1 mark)**

2 Calculate the pH of a solution that contains $2.5 \times 10^{-3} \, mol \, dm^{-3}$ $H^+(aq)$.   **(1 mark)**

pH of acids

Strong acids are fully dissociated in solution but weak acids are only partially dissociated in solution.

Strong acids

A **strong acid** is fully dissociated in solution.

For example, hydrochloric acid is a strong acid:

$HCl(aq) \rightarrow H^+(aq) + Cl^-(aq)$

full dissociation

When you calculate the pH of a strong acid, you assume:

- for a monobasic acid, $[H^+] = [acid]$

For example, for the pH of $0.250\,mol\,dm^{-3}$ HNO_3:

- $[H^+] = [HNO_3] = 0.250\,mol\,dm^{-3}$
- $pH = -\log_{10}(0.250)$
  $= 0.60$ to 2 decimal places

Acid dissociation constant, K_a

For a general weak acid, HA:

$HA(aq) \rightleftharpoons H^+(aq) + A^-(aq)$

You can write an expression for K_c but for acids this is called the **acid dissociation constant**:

$$K_a = \frac{[H^+(aq)][A^-(aq)]}{[HA(aq)]}$$

The lower the value of K_a, the weaker the acid.

Weak acids

A **weak acid** is only partially dissociated in solution. For example, ethanoic acid is a weak acid:

$CH_3COOH(aq) \rightleftharpoons CH_3COO^-(aq) + H^+(aq)$

shows partial dissociation

When you calculate the pH of a weak acid, you need to take into account the acid dissociation constant, K_a.

> **Maths skills** **pK_a**
>
> $K_a = -\log_{10}K_a$
>
> For example, K_a for propanoic acid is $1.35 \times 10^{-5}\,mol\,dm^{-3}$ at 298 K:
>
> $pK_a = -\log_{10}(1.35 \times 10^{-5}) = 4.87$
>
> The higher the pK_a is, the weaker the acid.

At equilibrium for an acid HA:

$[H^+(aq)] = [A^-(aq)]$

(because one HA molecule will dissociate to form one H^+ ion and one A^- ion).

This gives you two simplified equations that you can apply to weak acids in solution:

$$K_a = \frac{[H^+]^2}{[acid]} \qquad [H^+] = \sqrt{K_a \times [acid]}$$

Worked example

(a) Calculate the hydrogen ion concentration of $0.250\,mol\,dm^{-3}$ ethanoic acid.
$K_a = 1.74 \times 10^{-5}\,mol\,dm^{-3}$ at 298 K. **(1 mark)**

$$K_a = \frac{[H^+(aq)]^2}{[CH_3COOH(aq)]}$$

$$1.74 \times 10^{-5} = \frac{[H^+(aq)]^2}{0.250}$$

$[H^+]^2 = (1.74 \times 10^{-5}) \times 0.250 = 4.35 \times 10^{-6}$

$[H^+] = \sqrt{4.35 \times 10^{-6}} = 2.09 \times 10^{-3}\,mol\,dm^{-3}$

(b) Calculate the pH of the acid. **(1 mark)**

$pH = -\log_{10}(2.09 \times 10^{-3}) = -(-2.68)$
$= 2.68$

(c) State **two** assumptions made in your calculation in part (a). **(2 marks)**

1. The equilibrium concentration of the ethanoic acid is the same as its initial concentration.

2. The concentration of hydrogen ions due to the ionisation of water is negligible.

Notice that the pH of ethanoic acid (a weak acid) is greater than the pH of nitric acid (a strong acid) at the same concentration.

If you compare the values of K_a and of pK_a for ethanoic acid and propanoic acid, you should see that propanoic acid is the weaker of these two.

The first assumption is reasonable as long as the value for K_a is small (which it is here).

The second assumption is also reasonable, unless the acid is very dilute.

Now try this

Calculate the pH of these acids at 298 K, and express your answers to two decimal places:
(a) $0.500\,mol\,dm^{-3}$ hydrochloric acid **(1 mark)**
(b) $0.500\,mol\,dm^{-3}$ propanoic acid
    ($K_a = 1.35 \times 10^{-5}\,mol\,dm^{-3}$) **(2 marks)**

pH of bases

You can calculate the pH of a strong base if you take into account the ionic product of water, K_w.

Ionic product of water, K_w

Water reacts with itself in an acid–base reaction:

$$H_2O(l) + H_2O(l) \rightleftharpoons H_3O^+(aq) + OH^-(aq)$$

This can be simplified to:

$$H_2O(l) \rightleftharpoons H^+(aq) + OH^-(aq)$$

You can write an expression for K_c but for water this is called the **ionic product of water**:

$K_w = [H^+(aq)][OH^-(aq)]$ ——— [$H_2O(l)$] is a constant and is not included in the expression.

For pure water at 298 K:

- $K_w = 1.00 \times 10^{-14}\,mol^2\,dm^{-6}$

In the same way that $pK_a = -\log_{10}K_a$:

$$pK_w = -\log_{10}K_w$$

For pure water at 298 K,

- $pK_w = 14.0$

Maths skills   Neutral pH

In pure, neutral water, $[H^+(aq)] = [OH^-(aq)]$.

This means that:

- $K_w = [H^+(aq)]^2$
- $[H^+(aq)] = \sqrt{K_w}$

You can calculate the pH of water at 298 K:

$$pH = -\log_{10}(\sqrt{(1.00 \times 10^{-14})}) = 7.00$$

The dissociation of water is endothermic. This means that as the temperature increases:

- ☑ K_w increases
- ☑ pK_w decreases
- ☑   the pH of pure water decreases.

Neutral pH is only 7.00 at 298 K.

Worked example

(a) Explain why sodium hydroxide, NaOH, is a strong base. **(1 mark)**

It is fully dissociated in aqueous solution:

$$NaOH(aq) \rightarrow Na^+(aq) + OH^-(aq)$$

(b) Calculate the hydrogen ion concentration in 0.250 mol dm^{-3} sodium hydroxide at 298 K. ($K_w = 1.00 \times 10^{-14}\,mol^2\,dm^{-6}$) **(1 mark)**

$$K_w = [H^+(aq)][OH^-(aq)]$$

so $[H^+(aq)] = \dfrac{K_w}{[OH^-(aq)]}$

$$[H^+(aq)] = \frac{(1.00 \times 10^{-14})}{0.250}$$
$$= 4.00 \times 10^{-14}\,mol\,dm^{-3}$$

(c) Calculate the pH of this solution at 298 K. Express your answer to 2 decimal places. **(1 mark)**

$$pH = -\log_{10}(4.00 \times 10^{-14}) = -(-13.40)$$
$$= 13.40$$

Strong bases are fully dissociated in solution. They include KOH and Ca(OH)$_2$ (a dibasic base). Weak bases are only partially dissociated in solution. Ammonia is a weak base.

The temperature is quoted because the value for K_w varies with temperature.

When you calculate [OH$^-$(aq)]:

- [OH$^-$(aq)] = [monobasic strong base]
- [OH$^-$(aq)] = 2 × [dibasic strong base]

You may have to calculate the pH of a strong base from its concentration. The steps are:

1. Calculate [OH$^-$(aq)] from [base] (see above).
2. Calculate [H$^+$(aq)] from K_w and [OH$^-$]
3. Calculate pH using [H$^+$(aq)].

Enthalpy changes of neutralisation

Strong acids are fully dissociated in solution and have the greatest magnitude of $\Delta H^{\ominus}_{neut}$. Weak acids are partially dissociated in solution:

- Energy is needed to dissociate them.
- The magnitudes of their $\Delta H^{\ominus}_{neut}$ are lower.

Now try this

Calculate the pH of these strong bases at 298 K. Express your answers to one decimal place.

(a) 0.500 mol dm^{-3} sodium hydroxide. **(2 marks)**

(b) 0.500 mol dm^{-3} calcium hydroxide, Ca(OH)$_2$. **(2 marks)**

Buffer solutions

A **buffer** solution minimises the changes in pH when small amounts of acid or base are added to it.

Acidic buffers

Acidic buffer solutions have a pH < 7.

They are made by mixing a weak acid with a salt of its conjugate base. For example:

- ethanoic acid and sodium ethanoate.

An equilibrium forms:

$$CH_3COOH(aq) \rightleftharpoons CH_3COO^-(aq) + H^+(aq)$$

If a small amount of acid is added, H^+ ions react with CH_3COO^- ions to form CH_3COOH, and the position of equilibrium moves to the left.

If a small amount of base is added, OH^- ions react with two substances:

- with CH_3COOH (to form CH_3COO^- and H_2O)
- with H^+ (to form H_2O).

The position of equilibrium moves to the right.

Basic buffers

Basic buffer solutions have a pH > 7.

They are made by mixing a weak base with a salt of its conjugate acid. For example:

- ammonia solution and ammonium chloride.

An equilibrium forms:

$$NH_3(aq) + H_2O(l) \rightleftharpoons NH_4^+(aq) + OH^-(aq)$$

If a small amount of acid is added, H^+ ions react with two substances:

- with NH_3 (to form NH_4^+)
- with OH^- (to form H_2O).

The position of equilibrium moves to the right.

If a small amount of base is added, OH^- ions react with NH_4^+ ions, and the position of equilibrium moves to the left.

Controlling blood pH

The pH of arterial blood is maintained in the range 7.35 – 7.45 by buffers in the plasma.

The most important one is formed by:

- carbonic acid, H_2CO_3, a weak acid and
- hydrogencarbonate ions, HCO_3^-, a conjugate base.

An equilibrium forms:

$$H_2CO_3(aq) \rightleftharpoons HCO_3^-(aq) + H^+(aq)$$

- If $[H^+]$ goes up and pH falls, the position of equilibrium moves to the left.
- If $[H^+]$ goes down and pH rises, the position of equilibrium moves to the right.

Carbon dioxide and blood pH

Carbonic acid decomposes in aqueous solution:

$$H_2CO_3(aq) \rightleftharpoons H_2O(l) + CO_2(aq)$$

Dissolved carbon dioxide exists in equilibrium with gaseous carbon dioxide, which can be removed from the body via the lungs:

$$CO_2(aq) \rightleftharpoons CO_2(g)$$

If $[H^+]$ goes up and pH falls:

☑ $[H_2CO_3(aq)]$ increases

☑ $[CO_2(aq)]$ increases

☑ $[CO_2(g)]$ increases.

Excess CO_2 leaves the body via the lungs.

Worked example

0.5 dm³ of a buffer solution contains 0.100 mol of ethanoic acid ($K_a = 1.74 \times 10^{-5}$ mol dm⁻³) and 0.200 mol of sodium ethanoate. Estimate its pH.

(3 marks)

$[CH_3COOH] = 0.100 \div 0.5$
$= 0.200$ mol dm⁻³

$[CH_3COO^-] = 0.200 \div 0.5$
$= 0.400$ mol dm⁻³

$$[H^+] = K_a \times \frac{[CH_3COOH]}{[CH_3COO^-]}$$

$$[H^+] = 1.74 \times 10^{-5} \times \frac{0.200}{0.400} = 8.70 \times 10^{-6}$$

$pH = -\log_{10}[H^+]$

$pH = -\log_{10}(8.70 \times 10^{-6}) = 5.06$

The weak acid is only partially dissociated, so the calculation assumes that:

- $[CH_3COOH]$ at eqm = $[CH_3COOH]$ at start
- $[CH_3COO^-]$ at eqm = $[CH_3COONa]$ at start.

In general:

$$[H^+] = K_a \times \frac{[acid]}{[salt]}$$

$$pH = pK_a + \log_{10}\left(\frac{[salt]}{[acid]}\right)$$

Take care – the acid and salt are the other way up in this equation.

Now try this

Calculate the approximate pH of 0.25 dm³ of a buffer solution containing 0.10 mol of methanoic acid ($pK_a = 3.8$) and 0.10 mol of sodium methanoate. What do you notice? **(3 marks)**

More pH calculations

You can calculate the concentrations of solutions needed to produce a buffer with a given pH.

Worked example

A buffer solution, containing ethanoic acid and sodium ethanoate, is required with a pH of 5.06.

(a) Calculate its H^+ concentration. **(1 mark)**

$[H^+] = 10^{-pH}$
$[H^+] = 10^{-5.06} = 8.71 \times 10^{-6} \, mol \, dm^{-3}$

(b) Calculate the ratio $[CH_3COOH]/[CH_3COONa]$ required ($K_a = 1.74 \times 10^{-5} \, mol \, dm^{-3}$). **(2 marks)**

$$[H^+] = K_a \times \frac{[acid]}{[salt]}$$

$$\frac{[CH_3COOH]}{[CH_3COONa]} = \frac{[H^+]}{K_a} = \frac{8.71 \times 10^{-6}}{1.74 \times 10^{-5}} = 0.50$$

(c) $0.2 \, mol \, dm^{-3}$ ethanoic acid is supplied. Calculate the concentration of sodium ethanoate solution required when equal volumes are mixed. **(1 mark)**

$[CH_3COONa] = 0.2 \div 0.50 = 0.4 \, mol \, dm^{-3}$

Remember:
$pH = -\log_{10}[H^+]$
The value quoted for K_a is measured at 298 K. The calculation assumes that the buffer solution will be pH 5.06 at the same temperature.

If you are given the pK_a value of the weak acid, you could use this instead:
$$\frac{[CH_3COOH]}{[CH_3COONa]} = 10^{(pK_a - pH)}$$
For example, $pK_a = 4.76$
Ratio $[CH_3COOH]/[CH_3COONa] = 10^{(4.76 - 5.06)}$
$= 0.50$

The total volume will be double the volume of each individual component of the buffer.

In the buffer itself:
- $[CH_3COOH] = 0.1 \, mol \, dm^{-3}$
- $[CH_3COONa] = 0.2 \, mol \, dm^{-3}$

Diluting strong acids

Strong acids are fully dissociated in solution:
- $[H^+] = $ [monobasic acid]

For example, if you dilute $1.0 \, mol \, dm^{-3}$ hydrochloric acid:

$[HCl]/mol \, dm^{-3}$	Dilution factor	pH at 298 K
1.0	1×	0
0.1	10×	1
0.01	100×	2
0.001	1000×	3

The pH increases by 1 for each 10× dilution of a strong monobasic acid.

Diluting weak acids

Weak acids are partially dissociated in solution:
- $[H^+] = \sqrt{K_a \times [weak \, acid]}$

For example, if you dilute $1.0 \, mol \, dm^{-3}$ ethanoic acid ($K_a = 1.74 \times 10^{-5} \, mol \, dm^{-3}$ at 298 K):

$[CH_3COOH]/mol \, dm^{-3}$	Dilution factor	pH at 298 K
1.0	1×	2.38
0.1	10×	2.88
0.01	100×	3.38
0.001	1000×	3.88

The pH increases by 0.5 for each 10× dilution of a weak monobasic acid.

Practical skills — Diluting solutions

To achieve a 10 × dilution, you need to mix:
- 1 volume of the original solution with
- (10 − 1) = 9 volumes of water.
$50 \, cm^3$ of $1.0 \, mol \, dm^{-3}$ HCl, added to $450 \, cm^3$ of water, makes $500 \, cm^3$ of $0.1 \, mol \, dm^{-3}$ HCl.

Diluted buffers and pH

Factors affecting pH of an acidic buffer include:
- the value of K_a for the weak acid
- the ratio of [acid]/[salt].

These factors stay the same when a buffer is diluted. Unless diluted to a large extent, the pH of a buffer solution stays the same when diluted.

Now try this

In what proportions should $0.25 \, mol \, dm^{-3}$ $NH_4Cl(aq)$ and $0.25 \, mol \, dm^{-3}$ $NH_3(aq)$ be mixed to form a buffer at pH 9.65 at 298 K?   (K_a for $NH_4^+ = 5.62 \times 10^{-10} \, mol \, dm^{-3}$ at 298 K)   **(3 marks)**

Titration curves

A **titration curve** can show how the pH changes when monobasic acids and bases are mixed.

Strong acid into strong base

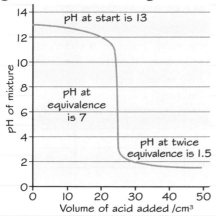

0.1 mol dm^{-3} HCl added from a burette to 25 cm^3 of 0.1 mol dm^{-3} NaOH in a conical flask.

Points on the curve

The **equivalence point** is the point at which the acid and base have been mixed in the exact proportions shown in the chemical equation. For monobasic acids and bases, this is when **equimolar** amounts have been mixed (same number of moles of acid and base).

On a titration curve, this is the pH halfway through the near vertical section.

The **end point** is the point in the titration where the indicator changes colour.

The equivalence point and end point are not the same as the so-called 'neutral point' – they do not necessarily happen at pH 7.

Strong acid into weak base

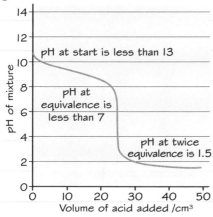

The pH falls rapidly at the start. The gradient then reduces until close to the equivalence point because a buffer solution forms.

Weak acid into weak base

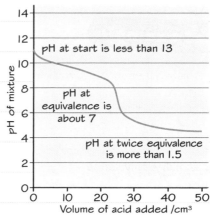

There may be no obvious change in pH at the equivalence point. Instead there may be a **point of inflexion** with a slight change in gradient.

Worked example

Sketch a titration curve for a weak acid added to a strong base. Indicate the pH at the start, equivalence, and twice equivalence. **(3 marks)**

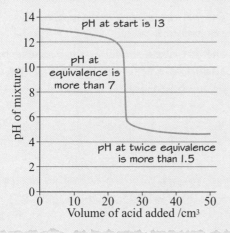

You need to be able to draw and interpret titration curves for **monobasic** acids and bases.

The answer assumes that 25 cm^3 of strong base is contained in the flask, and that the acid and base are at equal concentrations.

Notice that after equivalence, the gradient reduces because a buffer solution forms.

This consists of:
- the weak acid
- the salt of its conjugate base (formed during the titration).

Now try this

Sketch titration curves for the four combinations of weak or strong monobasic bases being added to weak or strong monobasic acids. **(12 marks)**

Determining K_a

For a strong base – weak acid titration, the pH at half-equivalence is equal to the pK_a of the acid.

Acid–base indicators

An acid–base **indicator** (HIn) is usually a weak acid. It has different colours depending on its pH. For example, methyl orange:

$$HIn(aq) \rightleftharpoons H^+(aq) + In^-(aq)$$

red   yellow

colour in acidic solutions   colour in alkaline solutions

The pH at which the indicator changes colour is determined by its equilibrium constant, K_{In}:

$$K_{In} = \frac{[H^+(aq)][In^-(aq)]}{[HIn(aq)]}$$

When $[HIn(aq)] = [In^-(aq)]$, methyl orange appears orange (a mixture of red and yellow):

$$K_{In} = [H^+(aq)] = 2.00 \times 10^{-4} \, mol \, dm^{-3}$$
$$pK_a = -\log_{10}(2.00 \times 10^{-4}) = 3.70$$

the pH at which colour changes

Choosing an indicator

Single indicators, such as methyl orange and phenolphthalein, provide a sharp end point. Their useful range is often just a few tenths of a pH unit either side of the pH at $[HIn] = [In^-(aq)]$.

Indicator	Methyl orange	Phenolphthalein
pK_{In}	3.70	9.30
pH range	3.10 – 4.40	8.30 – 10.0
HIn	red	colourless
In⁻	yellow	red

Suitable for strong acid – strong base and strong acid – weak base

Suitable for strong acid – strong base and weak acid – strong base

---

The graph shows a titration curve when $0.1 \, mol \, dm^{-3}$ sodium hydroxide solution is added to $25 \, cm^3$ of $0.1 \, mol \, dm^{-3}$ ethanoic acid. Use the graph to calculate the K_a of ethanoic acid. **(3 marks)**

Volume at equivalence = $25.0 \, cm^3$

Volume at half-equivalence = $\frac{25.0}{2} = 12.5 \, cm^3$

pH at half-equivalence = 4.8 (from graph)

At half-equivalence, $pK_a = pH = 4.8$

$K_a = 10^{-4.8} = 1.6 \times 10^{-5} \, mol \, dm^{-3}$

Do not assume that equivalence will be at $25.0 \, cm^3$ – always check the titration curve given.

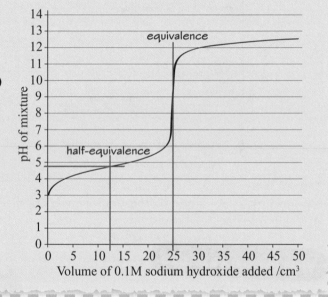

pH of mixture vs Volume of 0.1M sodium hydroxide added /cm³

Buffer action seen in a titration curve

In the titration curve above:

- The pH initially rises rapidly because a base is added to a weak acid.

- The gradient decreases because a buffer solution forms, consisting of ethanoic acid and sodium ethanoate, which resists changes in pH.

- The pH rises very rapidly near the equivalence point.

- The pH rises slowly again because base is added to a solution in which all the acid has been neutralised.

---

1   The K_{In} for phenol red is $1.2 \times 10^{-8} \, mol \, dm^{-3}$ at 298 K. Assuming that its colour changes begin when $[HIn] = 10 \times [In^-]$ and when $[HIn] = 0.1 \times [In^-]$, calculate its useful pH range. **(5 marks)**

2   The pH at half-equivalence in a titration involving sodium hydroxide solution and methanoic acid is 3.8. Calculate K_a for methanoic acid. **(1 mark)**

Exam skills 8

This exam-style question uses knowledge and skills you have already revised. Have a look at pages 80–84 for a reminder about **pH** calculations and **buffers**.

Worked example

Ethanoic acid, CH_3COOH, is a weak acid found in vinegar. A titration was carried out in which $0.200 \, mol \, dm^{-3}$ sodium hydroxide solution was added from a burette to $25.0 \, cm^3$ of $0.192 \, mol \, dm^{-3}$ dilute ethanoic acid in a conical flask.

(a) Calculate the volume of sodium hydroxide solution added at the equivalence point.

(2 marks)

Amount of $CH_3COOH = 0.192 \times (25.0 \times 10^{-3})$
$$= 4.80 \times 10^{-3} \, mol$$
Amount of $NaOH = 4.80 \times 10^{-3} \, mol$
$$Volume \ of \ NaOH = \frac{(4.80 \times 10^{-3})}{0.200}$$
$$= 0.24 \, dm^3 = 24.0 \, cm^3$$

(b) Calculate the pH of $0.200 \, mol \, dm^{-3}$ sodium hydroxide at 298 K.
($K_w = 1.00 \times 10^{-14} \, mol^2 \, dm^{-6}$)   **(2 marks)**
$$[H^+] = \frac{K_w}{[NaOH]} = \frac{(1.00 \times 10^{-14})}{0.200}$$
$$= 5.00 \times 10^{-14} \, mol \, dm^{-3}$$
$$pH = -log_{10}[H^+] = -log_{10}(5.00 \times 10^{-14})$$
$$= 13.3$$

(c) Calculate the pH of the ethanoic acid before any sodium hydroxide has been added. ($K_a = 1.74 \times 10^{-5} \, mol \, dm^{-3}$ at 298 K)   **(2 marks)**
$$[H^+] = \sqrt{1.74 \times 10^{-5} \times 0.192}$$
$$= 1.828 \times 10^{-3} \, mol \, dm^{-3}$$
$$pH = 2.74$$

(d) Calculate the pH of the solution in the conical flask when $10.0 \, cm^3$ of NaOH(aq) has been added.   **(6 marks)**

Amount of NaOH added $= 0.200 \times (10.0 \times 10^{-3})$
$$= 2.00 \times 10^{-3} \, mol$$
Amount of CH_3COONa formed $= 2.00 \times 10^{-3} \, mol$
Amount of CH_3COOH at start $= 4.80 \times 10^{-3} \, mol$
Amount of CH_3COOH left $= (4.80 \times 10^{-3}) - (2.00 \times 10^{-3})$
$$= 2.80 \times 10^{-3} \, mol$$
$$[H^+] = K_a \times \frac{[CH_3COOH]}{[CH_3COONa]}$$
$$= 1.74 \times 10^{-5} \times \frac{2.80 \times 10^{-3}/volume}{2.00 \times 10^{-3}/volume}$$
$$= 2.436 \times 10^{-5} \, mol \, dm^{-3}$$
$$pH = 4.61$$

There is a 1:1 ratio between ethanoic acid and sodium hydroxide in the equation:
$$NaOH + CH_3COOH \rightarrow CH_3COONa + H_2O$$
Remember to convert cm^3 to dm^3:
• divide by 1000, or
• multiply by 10^{-3} (as here).
Check at each step that you use the correct values given in the question.

Maths skills The values given in the question are expressed to 3 significant figures, so you should express your answers to 3 significant figures unless told otherwise.
If a final answer depends on a value you calculate, it makes sense either to:
• put the intermediate value into your calculator's memory and use that precise value in your calculation, or,
• write down the intermediate value to one more significant figure than required, and use that precise value.

$$[H^+] = \sqrt{K_a \times [CH_3COOH]}$$
This assumes:
1. $[CH_3COOH]$ at equilibrium $= [CH_3COOH]$ at the start.

2. $[H^+]$ due to the ionisation of water is negligible.
You could be asked for these assumptions in a question.

Some of the ethanoic acid has reacted with the sodium hydroxide, forming sodium ethanoate and water.

This means that the mixture in the conical flask is an acidic buffer involving ethanoic acid and ethanoate ions.

So, the pH calculation is for a buffer solution, even though the question does not directly tell you this: you would obtain an incorrect pH if you assumed that only ethanoic acid was present.

Remember that at half-equivalence, $[H^+] = K_a$ (and pH = pK_a). In this example, pH = 4.76 when $12.0 \, cm^3$ of NaOH is added.

Born–Haber cycles 1

Born–Haber cycles are used to calculate enthalpy changes involving ionic compounds.

Lattice enthalpy vs bond enthalpy

Bond strength in covalent compounds is given by the **bond enthalpy** (see page 11):

- the enthalpy change when one mole of a stated covalent bond in the gaseous state is broken.

Bond strength in ionic compounds is given by the lattice energy (also sometimes called the lattice enthalpy). **Lattice energy** is:

- the energy change when one mole of an ionic solid is formed from its gaseous ions.

$\Delta H_{lat}^{\ominus}$ (standard lattice enthalpy) is measured at 100 kPa and a stated temperature, usually 298 K. For example, for sodium chloride:

$$Na^+(g) + Cl^-(g) \rightarrow NaCl(s)$$
$$\Delta H_{lat}^{\ominus} = -780 \text{ kJ mol}^{-1}$$

Electron affinity

Ionisation energies, E_m, refer to the formation of positively charged ions (cations; see page 6).

Electron affinities, E_{aff}, refer to the formation of negatively charged ions (anions).

Electron affinity is the energy change when one mole of electrons is gained by one mole of gaseous atoms or anions to form one mole of gaseous anions. For example:

- first electron affinity of chlorine:
$$Cl(g) + e^- \rightarrow Cl^-(g) \quad E_{aff1} = -349 \text{ kJ mol}^{-1}$$

- first electron affinity of oxygen:
$$O(g) + e^- \rightarrow O^-(g) \quad E_{aff1} = -141 \text{ kJ mol}^{-1}$$

- second electron affinity of oxygen:
$$O^-(g) + e^- \rightarrow O^{2-}(g) \quad E_{aff2} = +798 \text{ kJ mol}^{-1}$$

Enthalpy change of atomisation

Enthalpy change of atomisation, $\Delta_{at}H$, is:

- the enthalpy change when one mole of <u>gaseous</u> atoms are formed

- from an element in its standard state.

$\Delta_{at}H^{\ominus}$ (standard enthalpy change of atomisation) is measured at 100 kPa and a stated temperature, usually 298 K. For example, for chlorine:

$$\tfrac{1}{2}Cl_2(g) \rightarrow Cl(g) \quad \Delta_{at}H^{\ominus}[\tfrac{1}{2}Cl_2(g)] = +121 \text{ kJ mol}^{-1}$$

Endothermic (bonds broken)

Worked example

Fill in each enthalpy change on the Born–Haber cycle below. **(6 marks)**

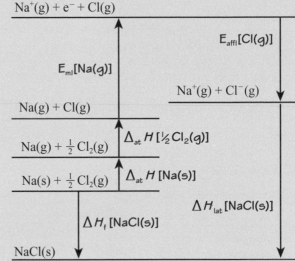

Note the use of state symbols, and square brackets. The enthalpy level increases upwards:
- endothermic changes have up arrows
- exothermic changes have down arrows.

Born–Haber cycles as enthalpy cycles

You may also see Born–Haber cycles drawn in a similar way to enthalpy cycles (see page 68).

Here is the cycle in the Worked Example drawn this way:

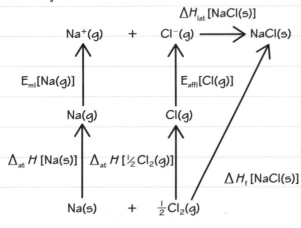

Now try this

Draw and label a Born–Haber cycle for the formation of potassium bromide, KBr.

(12 marks)

It will be similar to either of the ones above, but for K(s) and $\tfrac{1}{2}Br_2(l)$.

Born–Haber cycles 2

You can calculate enthalpy changes for ionic compounds using Born–Haber cycles and given data.

First law of thermodynamics

Enthalpy changes for covalent compounds are calculated by applying Hess's Law (page 68), which relies on the first law of thermodynamics.

In the same way, calculations using Born–Haber cycles for ionic compounds rely on this principle:

- the total enthalpy change for a reaction is independent of the pathway taken.

Positive values for E_{aff}

Values for the first electron affinity, E_{aff1}, are negative for almost all elements.

They are positive for the noble gases due to repulsion from the electrons in the outer shell.

Values for the second electron affinity, E_{aff2}, also tend to be positive, due to repulsion between the electron and the anion.

Worked example

(a) Use the Born–Haber cycle and the data in this table to calculate the enthalpy change of formation of sodium chloride, NaCl(s). **(2 marks)**

Energy change	$\Delta H/\text{kJ mol}^{-1}$
Enthalpy change of atomisation of sodium	+107
Enthalpy change of atomisation of chlorine	+122
First ionisation energy of sodium	+496
First electron affinity of chlorine	−349
Lattice energy for sodium chloride	−780

$\Delta H_f[\text{NaCl(s)}]$
$= +107 + 496 + 122 + (−349) + (−780)$
$= 725 − 349 − 780 = −404 \text{ kJ mol}^{-1}$

(b) Use the Born–Haber cycle, and data from the two tables, to calculate the lattice energy of sodium oxide, $\text{Na}_2\text{O(s)}$. **(2 marks)**

Energy change	$\Delta H/\text{kJ mol}^{-1}$
Enthalpy change of atomisation of oxygen	+249
First electron affinity of oxygen	−141
Second electron affinity of oxygen	+798
Enthalpy change of formation of sodium oxide	−414

$\Delta H_{lat}[\text{Na}_2\text{O(s)}]$
$= −(798) − (−141) − (992) − (249) − (214) +$
$(−414)$
$= −798 + 141 − 992 − 249 − 214 − 414$
$= −2526 \text{ kJ mol}^{-1}$

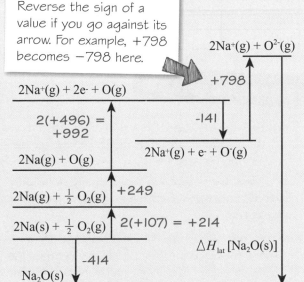

It can help if you write the values for each enthalpy change onto the Born–Haber cycle.

Reverse the sign of a value if you go against its arrow. For example, +798 becomes −798 here.

Now try this

(a) Draw a Born–Haber cycle for the formation of magnesium chloride, $\text{MgCl}_2\text{(s)}$, showing the species and their state symbols at each stage. **(4 marks)**

(b) Calculate the lattice energy of magnesium chloride using data above and on the right. **(2 marks)**

Energy change	$\Delta H/\text{kJ mol}^{-1}$
$\Delta_{at}H[\text{Mg(s)}]$	+148
$E_{m1}[\text{Mg(g)}]$	+738
$E_{m2}[\text{Mg(g)}]$	+1451
$\Delta H_f[\text{MgCl}_2\text{(s)}]$	−641

An ionic model

Lattice energies can be calculated from a model of ionic bonding. A degree of covalent bonding is revealed when these are compared with values obtained experimentally using Born–Haber cycles.

Cations and polarising power

The cations (positive ions) in an ionic lattice can attract electrons in the anions, distorting the anion's even distribution of electron density.

The **polarising power** of a cation:

- increases with increasing charge
- increases with decreasing ionic radius.

For example, Al^{3+} ions have a higher polarising power than Na^+ ions because they have:
- a higher charge (3+ vs 1+)
- a smaller ionic radius (0.053 nm vs 0.102 nm).

Anions and polarisation

The even distribution of electron density of the anions (negative ions) in an ionic lattice can be distorted or polarised by the cations.

The **polarisability** of an anion:

- increases with increasing charge
- increases with increasing ionic radius.

For example, Cl^- ions are more easily polarised than F^- ions. They have the same charge but:
- a larger ionic radius (0.180 nm vs 0.133 nm).

Polarisation of anions leads to a **degree of covalent bonding** in ionic compounds.

Covalent character in ionic compounds

Ionic compounds have covalent character due to polarisation of their anions:

- The greater the degree of covalent bonding, the larger the percentage difference between the experimental and theoretical lattice energies.

For example, for three Group I halides:

Compound	% difference
AgCl	7.96
NaCl	1.28
NaI	2.55

—Ag^+ is smaller than Na^+.

—I^- is larger than Cl^-.

Factors in the ionic model

Theoretical lattice energies can be calculated using electrostatic theory and a model of ionic bonding. The factors involved include:
- the charge on an electron (a constant)
- the number of charges on the ions
- distances between the centres of the ions.

The ionic model assumes:

☑ The ions are perfect spheres.

☑ Their charge is distributed evenly.

☑ The ions pack together in a regular way.

If the actual situation is different, it leads to a difference between the theoretical and experimental values for lattice energy.

Worked example

(a) Explain why there is a greater difference between the experimental and theoretical lattice energies for CaS than for CaO. **(2 marks)**

S^{2-} ions are larger and more easily polarised than O^{2-} ions, so CaS has a greater covalent character.

(b) Predict whether the lattice energy of calcium oxide, CaO, is more or less exothermic than the lattice energy of calcium sulfide, CaS. **(4 marks)**

The lattice energy of CaO will be more exothermic than CaS. This is because S^{2-} ions and O^{2-} have the same charge, but S^{2-} ions are larger. This means that O^{2-} ions form stronger attractions than S^{2-} ions.

Notice that the answer echoes the question; it does not say that the lattice enthalpy of CaO is 'larger' or 'higher' than that of CaS.

The differences are:
- CaO, 0.38%
- CaS, 0.83%

So both have very little covalent character compared to the Group I halides (table, above left).

The experimental lattice energies for these two compounds are:
- CaO, $-3401 \, kJ \, mol^{-1}$
- CaS, $-3013 \, kJ \, mol^{-1}$

Now try this

Predict whether the experimental lattice energy of lithium bromide, LiBr, will be more or less negative than its theoretical lattice energy. Explain your answer. **(3 marks)**

Dissolving

Energy cycles and energy level diagrams can be used to carry out calculations involving dissolving.

Enthalpy change of solution

The **enthalpy change of solution**, $\Delta_{sol}H$, is:

- the enthalpy change when one mole of an ionic solid
- dissolves in excess water to form an infinitely dilute solution.

Values for $\Delta_{sol}H$ can be positive or negative.

For example:

- $NaCl(s) + aq \rightarrow Na^+(aq) + Cl^-(aq)$
  $\Delta_{sol}H = +3.9 \, kJ \, mol^{-1}$ (temperature decreases)
- $CaCl_2(s) + aq \rightarrow Ca^{2+}(aq) + 2Cl^-(aq)$
  $\Delta_{sol}H_s = -82 \, kJ \, mol^{-1}$ (temperature increases)

Enthalpies of solution can be determined experimentally using calorimetry (see page 66).

Enthalpy change of hydration

The **enthalpy change of hydration**, $\Delta_{hyd}H$, is:

- the enthalpy change when one mole of gaseous ions dissolves
- in excess water to form aqueous ions.

Values for $\Delta_{hyd}H$ are always negative (because bonds form between ions and water molecules).

For example:

- $Na^+(g) + aq \rightarrow Na^+(aq)$ $\Delta_{hyd}H = -406 \, kJ \, mol^{-1}$
- $Cl^-(g) + aq \rightarrow Cl^-(aq)$ $\Delta_{hyd}H = -363 kJ \, mol^{-1}$
- $Ca^{2+}(g) + aq \rightarrow Ca^{2+}(aq)$
  $\Delta_{hyd}H = -1653 \, kJ \, mol^{-1}$

aq represents water here

Factors affecting $\Delta_{hyd}H$

Enthalpy change of hydration depends upon:

- ionic charge
- ionic radius

$\Delta_{hyd}H$ becomes:

- more negative as the charge increases (see Na^+ and Ca^{2+} above).
- less negative as the ionic radius increases (such as going down a group).

A small, highly charged ion has stronger electrostatic forces of attraction with water molecules than a large ion with a lower charge. Page 16 gives you more information about these forces of attraction.

Enthalpy level diagrams for dissolving

You can make energy level diagrams for dissolving:

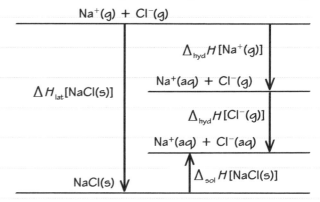

Worked example

(a) Draw an enthalpy cycle diagram to represent the enthalpy changes that occur when calcium chloride, $CaCl_2$, dissolves in water. **(2 marks)**

(b) Calculate the lattice energy for calcium chloride using data supplied on this page. **(2 marks)**

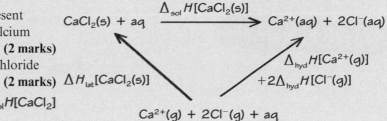

$\Delta H_{lat} = \Delta_{hyd}H[Ca^{2+}] + 2\Delta_{hyd}H[Cl^-] - \Delta_{sol}H[CaCl_2]$

$\Delta H_{lat} = -1653 + 2(-338) - (-82)$

$\Delta H_{lat} = -1653 - 676 + 82 = -2247 \, kJ \, mol^{-1}$

You use $2\Delta_{hyd}H[Cl^-]$, not just $\Delta_{hyd}H[Cl^-]$, because there are two chloride ions in the formula for calcium chloride.

Remember that lattice energies are negative, so check your answer to see if it makes sense.

Now try this

The lattice energy for calcium fluoride, CaF_2, is $-2630 \, kJ \, mol^{-1}$ and the enthalpy change of hydration of fluoride ions is $-474 \, kJ \, mol^{-1}$. Calculate the enthalpy change of solution for calcium fluoride. **(3 marks)**

Use the enthalpy change of hydration for Ca^{2+} ions on this page.

Entropy

Entropy changes occur during changes of state, dissolving and chemical reaction.

Entropy

Entropy, S, is a measure of the **disorder** of a system. Factors that affect entropy include:

- the number of particles present
- the complexity of the particles
- how many ways the particles can be arranged.
  ↘ related to state and temperature

Entropy values can be determined, whereas enthalpy values cannot (only enthalpy changes).

Entropy and water

Taking water as an example, you can see that its different states have different entropy values:

Substance	Entropy, S/J K^{-1}mol^{-1}	Disorder
$H_2O(s)$	62.1	least disordered
$H_2O(l)$	69.9	↓
$H_2O(g)$	188.7	most disordered

The units are <u>not</u> kJ mol^{-1}

Entropy changes happen during changes of state

In a solid, the number of ways the particles can be arranged is very limited.

In a gas, there are very many ways that the particles can be arranged.

State	solid	liquid	gas
Arrangement	ordered	random	random
Movement	vibrate about fixed positions	move around each other	move rapidly in any direction
State change	**Melting** some bonds break increase in particle movement increase in disorder increase in entropy		**Boiling** remaining bonds break big increase in particle movement big increase in disorder big increase in entropy

Worked example

(a) Explain why the entropy of the system increases when sodium chloride dissolves in water.

(1 mark)

The ions are arranged more randomly in the solution than they are in the solid, so there is a greater number of ways to arrange the particles.

(b) Predict the change in entropy in the system in this reaction: $Ni(s) + 4CO(g) \rightarrow Ni(CO)_4(g)$. Justify your answer. **(1 mark)**

The entropy of the system decreases because there are fewer moles of product than there are moles of reactant, so there are fewer ways to arrange the particles.

The **system** refers to the sodium chloride, water and sodium chloride solution.

The **surroundings** would refer to everything else.

The system refers to the reactants and products here.

Now try this

1 Explain the change in entropy in the system when water freezes. **(3 marks)**

2 Predict the change in entropy in the system in this reaction: $N_2O_4(g) \rightarrow 2NO_2(g)$. Justify your answer. **(1 mark)**

Calculating entropy changes

You can calculate entropy changes if you know the entropies of the reactants and products.

Entropy change in a system

You can calculate the entropy change in a system, $\Delta S^{\ominus}_{system}$, using this expression:

$$\Delta S^{\ominus}_{system} = \Sigma S^{\ominus}_{products} - \Sigma S^{\ominus}_{reactants}$$

'Sum of'

For example, calcium carbonate decomposes when heated strongly:

$$CaCO_3(s) \rightarrow CaO(s) + CO_2(g)$$

$S^{\ominus}/J\,K^{-1}\,mol^{-1}$ +92.9   +39.7   +213.6

$\Delta S^{\ominus}_{system} = (+39.7 + 213.6) - (+92.9)$

$= +160.4\,J\,K^{-1}\,mol^{-1}$

Total entropy change in a reaction

You can calculate the total entropy change in any reaction using this expression:

$$\Delta S^{\ominus}_{total} = \Sigma S^{\ominus}_{system} + \Sigma S^{\ominus}_{surroundings}$$

Using the calcium carbonate example at 890 K:

$\Delta S^{\ominus}_{total} = +160.4 + (-200)$

$= -39.6\,J\,K^{-1}\,mol^{-1}$

Entropy change in the surroundings

To calculate $\Delta S^{\ominus}_{surroundings}$ you need to know:

 the enthalpy change, $\Delta H^{\ominus}$

$$\Delta H^{\ominus} = \Sigma \Delta H^{\ominus}_f [products] - \Sigma \Delta H^{\ominus}_f [reactants]$$

 the temperature, T.

You then use this expression:

$$\Delta S^{\ominus}_{surroundings} = -\frac{\Delta H^{\ominus}}{T}$$

Maths skills $\Delta H^{\ominus}$ is measured in $kJ\,mol^{-1}$, so you must multiply its value by 1000 to convert it to $J\,mol^{-1}$ before using the expression.

Using the calcium carbonate example at 890 K:

$$CaCO_3(s) \rightarrow CaO(s) + CO_2(g)$$

$\Delta H^{\ominus}_f/kJ\,mol^{-1}$ −1207   −635   −394

$\Delta H^{\ominus} = -635 + (-394) - (-1207) = +178\,kJ\,mol^{-1}$

$\Delta H^{\ominus} = +178 \times 1000 = +178\,000\,J\,mol^{-1}$

$\Delta S^{\ominus}_{surroundings} = \underline{-178\,000 \div 890}$

$= -200\,J\,K^{-1}\,mol^{-1}$

Remember: minus $\frac{\Delta H^{\ominus}}{T}$

Gibbs energy

The **feasibility** of a reaction is determined by:

• the balance between the entropy change and the enthalpy change.

It can be represented by this expression:

$$\Delta G^{\ominus} = \Delta H^{\ominus} - T\Delta S^{\ominus}_{system}$$

Gibbs energy / $kJ\,mol^{-1}$

A **feasible reaction** is a reaction that is possible at a given temperature when the entropy change and enthalpy change are taken into account.

A reaction is feasible if $\Delta S^{\ominus}_{total}$ is positive.

In the calcium carbonate example above, $\Delta S^{\ominus}_{total}$ is negative – the reaction is not feasible at 890 K.

Worked example

Calculate the temperature at which the thermal decomposition of calcium carbonate is feasible.

$\Delta H^{\ominus} = +178\,kJ\,mol^{-1}$

$\Delta S^{\ominus}_{system} = +160.4\,J\,K^{-1}\,mol^{-1}$ **(2 marks)**

$\Delta G^{\ominus} = \Delta H^{\ominus} - T\Delta S^{\ominus}_{system}$

The reaction becomes feasible when $\Delta G = 0$

Therefore: $T = \dfrac{\Delta H^{\ominus}}{\Delta S^{\ominus}_{system}}$

$\Delta H^{\ominus} = 178\,000\,J\,mol^{-1}$

$T = 178\,000 \div 160.4 = 1110\,K$

Maths skills If $\Delta G^{\ominus} = \Delta H^{\ominus} - T\Delta S^{\ominus}_{system}$ and $\Delta G^{\ominus} = 0$

$\Delta H^{\ominus} - T\Delta S^{\ominus}_{system} = 0$

$-T\Delta S^{\ominus}_{system} = -\Delta H^{\ominus}$

$T = \dfrac{\Delta H^{\ominus}}{\Delta S^{\ominus}_{system}}$

A reaction is only feasible if $\Delta G^{\ominus} \leq 0$ so the temperature is calculated for $\Delta G^{\ominus} = 0$.

The calculation assumes that ΔH does not change with temperature. It shows that calcium carbonate is thermodynamically stable at temperatures below 1110 K.

Now try this

Use the data in the table to determine the temperatures at which these changes become feasible:

(a) $H_2O(l) \rightarrow H_2O(g)$.   **(4 marks)**

(b) $H_2O(l) \rightarrow H_2O(s)$.   **(4 marks)**

	$\Delta H^{\ominus}_f/kJ\,mol^{-1}$	$S^{\ominus}/J\,K^{-1}\,mol^{-1}$
$H_2O(s)$	−291.8	+48.0
$H_2O(l)$	−285.8	+69.9
$H_2O(g)$	−241.8	+188.7

Gibbs energy and equilibrium

Feasible reactions may not occur because they have high activation energies.

Feasible reactions and kinetic factors

A reaction is thermodynamically feasible at a given temperature when $\Delta G \leqslant 0$ (see page 93).

However, in practice a feasible reaction may not happen because:

- its **activation energy** is very high.

For example, the combustion of hydrogen:

$$H_2(g) + \tfrac{1}{2}O_2(g) \rightarrow H_2O(l)$$

$\Delta H^\ominus = -286\,kJ\,mol^{-1}$ (highly exothermic)

$\Delta G^\ominus = -272\,kJ\,mol^{-1}$ (feasible)

The activation energy is about $+71\,kJ\,mol^{-1}$, so the reaction is not **feasible** at 298 K. To start:
- the temperature must be increased, or
- a catalyst such as platinum must be used.

'Non-feasible' reactions that may happen

If you calculate $\Delta G^\ominus$ for a reaction, the value is valid for the stated standard conditions, usually:

- 298 K
- 100 kPa pressure
- solutions at $1\,mol\,dm^{-3}$

Some non-feasible reactions become feasible under non-standard conditions. For example, this reaction is used to make chlorine in the lab:

$$4HCl(aq) + MnO_2(s)$$
$$\rightarrow MnCl_2(aq) + 2H_2O(l) + Cl_2(g)$$

$\Delta G^\ominus$ is positive, but ΔG is negative when $10\,mol\,dm^{-3}$ hydrochloric acid is used.

Reversible reactions

This expression links Gibbs energy, ΔG, and the equilibrium constant K for a reversible reaction:

$$\Delta G = -RT \ln K$$

ΔG is Gibbs energy/$J\,mol^{-1}$ (*not* $kJ\,mol^{-1}$)

R is the gas constant ($8.31\,J\,mol^{-1}\,K^{-1}$)

T is the absolute temperature (K)

$\ln K$ is the natural logarithm of K (no units)

 Maths skills Natural logarithms are logarithms to the base e, $\log_e$. Your calculator will have a button marked ln (the one marked log is for $\log_{10}$). For example:

$\ln(10) = 2.30$ – check you get this answer.

ΔG and K

You can calculate the **thermodynamic equilibrium constant**, K, by rearranging the expression opposite like this:

$$K = e\left(-\frac{\Delta G}{RT}\right)$$

Value of ΔG	Value of K
large negative	>1 and very large
0	1
large positive	<1 and very small

☑ feasible reactions in terms of ΔG have large values for K

☑ reactions with large values for K are feasible in terms of ΔG

Worked example

Calcium carbonate decomposes at 1200 K:

$$CaCO_3(s) \rightleftharpoons CaO(s) + CO_2(g)$$

At this temperature, $\Delta G = -13.8\,kJ\,mol^{-1}$.

(a) Calculate the value for the equilibrium constant, K, at 1200 K. **(2 marks)**

$-\Delta G/RT = -(-13.8 \times 1000)/(8.31 \times 1200)$
$\qquad\qquad = 1.384$

$K = e^{1.384} = 3.99$ (no units)

(b) Comment on your answer to part (a). **(2 marks)**

The value for K is more than 1, so the position of equilibrium lies to the right, favouring products.

The first line in the calculation gives the answer to one more significant figure than the final answer. This reduces rounding errors.

Now try this

Ammonia is manufactured from nitrogen and hydrogen: $N_2(g) + 3H_2(g) \rightleftharpoons 2NH_3(g)$
The process is carried out at about 500 K, at which $\Delta G = +13.9\,kJ\,mol^{-1}$ (at 100 kPa).

(a) Calculate the value for the equilibrium constant, K, at 500 K. **(2 marks)**

(b) Comment on your answer to part (a). **(2 marks)**

Redox and standard electrode potential

The electrode potential of a metal or non-metal in contact with its aqueous ions can be measured.

Reduction

Reduction can be defined as:
- loss of oxygen
- gain of lectrons
- decrease in oxidation number.

You can see how to calculate oxidation number using assigned oxidation numbers on page 20.

Disproportionation

Disproportionation occurs when an element in a single species is simultaneously oxidised and reduced. For example:

$$Cu_2O(s) + H_2SO_4(aq)$$
$$\rightarrow Cu(s) + CuSO_4(aq) + H_2O(l)$$

- copper(I) is reduced to copper: $+1 \rightarrow 0$
- copper(I) is oxidised to copper(II): $+1 \rightarrow +2$

Oxidation

Oxidation can be defined as:
- gain of oxygen
- loss of electrons
- increase in oxidation number.

Redox

Page 21 describes the idea of **redox**, and page 22 explains how to construct ionic equations.

For example, manganate(VII) ions can oxidise iron(II) ions:

$$MnO_4^- + 8H^+ + 5Fe^{2+}$$
$$\rightarrow Mn^{2+} + 4H_2O + 5Fe^{3+}$$

- manganate(VII) is reduced to manganese(II)
- iron(II) is oxidised to iron(III)

Manganate(VII) is acting as an **oxidising agent** and iron(II) is acting as a **reducing agent** here.

Electrode potential

When a metal is placed in a solution of its ions, an equilibrium is eventually established.

For example: $Mg^{2+}(aq) + 2e^- \rightleftharpoons Mg(s)$

An electrical potential, E, exists between the metal and the solution, which form a **half-cell**.

This electrode potential:

✗ cannot be measured directly

✓ must be measured against a reference electrode (e.g. the standard hydrogen electrode).

Standard electrode potential, $E^{\ominus}$, of a half-cell is:

- the electromotive force, emf, of a cell
- where the standard hydrogen electrode is on the left and the half-cell is on the right, at
- 298 K temperature
- 100 kPa pressure of gases
- 1.00 mol dm^{-3} concentration of ions.

The emf is the potential difference (voltage) measured using a high-resistance voltmeter.

Worked example

Describe the essential features of the standard hydrogen electrode.   **(4 marks)**

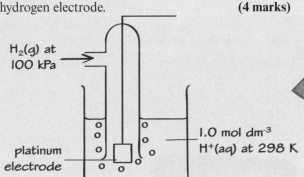

H$_2$(g) at 100 kPa

platinum electrode

1.0 mol dm^{-3} H$^+$(aq) at 298 K

The aqueous hydrogen ions can be provided by 1.0 mol dm^{-3} hydrochloric acid.

An equilibrium is established:

$$H^+(aq) + e^- \rightleftharpoons \tfrac{1}{2}H_2(g)$$

By definition, $E^{\ominus} = 0\,V$ for this electrode.

Platinum is an unreactive metal. It is present as porous platinum ('platinum black'), which:
- provides a large surface area for the reaction
- allows the conduction of electrons.

Now try this

(a) Describe, without the use of a diagram, the essential features of the standard hydrogen electrode. Include the conditions needed and the equation for the reaction.   **(5 marks)**

(b) Why does $E^{\ominus} = 0\,V$ for the standard hydrogen electrode?   **(1 mark)**

Measuring standard emf

The standard hydrogen electrode can be used to measure $E^\ominus$ values, and $E^\ominus_{cell}$ values can be calculated from the $E^\ominus$ values for two half-cells.

Combining two half-cells

Two half-cells are combined to make an **electrochemical cell**.

Two connections between them are needed:

1 An external electrical circuit, connecting the two electrodes and allowing electrons to flow from one half-cell to the other.

2 A **salt bridge**, connecting the two aqueous solutions and allowing ions to move between the two half-cells.

A salt bridge can be a strip of filter paper soaked in concentrated **potassium nitrate** solution.

This is chosen to avoid the possibility of a precipitate forming with ions from the half-cells.

Conventional representation

Half-cells are represented using **cell diagrams**.

Magnesium electrode in magnesium ions:
$$Mg^{2+}(aq)\,|\,Mg(s) \quad \text{phase boundary}$$

Standard hydrogen electrode:
$$H^+(aq)\,|\,H_2(g)\,|\,Pt(s) \quad \text{the platinum electrode}$$

Platinum electrode in a mixture of iron ions:
$$Fe^{3+}(aq),Fe^{2+}(aq)\,|\,Pt(s)$$
separates species in the same phase

Cells are also represented this way. For example, for a standard hydrogen electrode on the left and an $Mg^{2+}(aq)\,|\,Mg(s)$ half-cell on the right:
$$Pt(s)\,|\,H_2(g)\,|\,H^+(aq) \vdots\vdots Mg^{2+}(aq)\,|\,Mg(s) \quad \text{the salt bridge}$$

Measuring electrode potential, $E^\ominus$

Under standard conditions:

- emf of a cell, $E^\ominus_{cell} = E^\ominus_{right} - E^\ominus_{left}$

The standard electrode potential, $E^\ominus$, for the standard hydrogen electrode is $0\,V$.

If this electrode is placed on the left:

$$E^\ominus_{cell} = E^\ominus_{right} - 0 = E^\ominus_{right}$$

For example, if the cell in the box (top right) is set up, the reading on the high-resistance voltmeter will be $-2.37\,V$.

This means that $E^\ominus$ for $Mg^{2+}(aq)\,|\,Mg(s) = -2.37\,V$

Measuring standard emf, $E^\ominus_{cell}$

The $E^\ominus_{cell}$ value for a cell is measured by:

- constructing two half-cells
- connecting them using a salt bridge and external circuit
- ensuring standard conditions
- measuring the potential using a high-resistance voltmeter (see page 98).

Note that, if you know the value for $E^\ominus$ for one of the half-cells, with $E^\ominus_{cell}$ you can calculate the $E^\ominus$ for the other half-cell.

Worked example

This electrochemical cell was set up under standard conditions. Calculate its $E^\ominus_{cell}$. **(1 mark)**

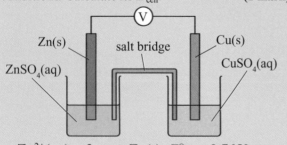

$$Zn^{2+}(aq) + 2e^- \rightleftharpoons Zn(s), E^\ominus = -0.76\,V$$
$$Cu^{2+}(aq) + 2e^- \rightleftharpoons Cu(s), E^\ominus = +0.34\,V$$

$$E^\ominus_{cell} = E^\ominus_{right} - E^\ominus_{left}$$
$$= +0.34 - (-0.76) = +1.10\,V$$

Apart from the standard electrode potential of the standard hydrogen electrode, you do not need to recall $E^\ominus$ values.

Notice that, by convention, the half-equations are written as **reduction reactions**.

Now try this

1. Write the cell diagram for the cell needed to measure $E^\ominus_{cell}$ for the mixture of iron ions shown at the top of the page. **(1 mark)**

2. Calculate $E^\ominus_{cell}$ for this cell:
$$Mg(s)\,|\,Mg^{2+}(aq) \vdots\vdots Zn^{2+}(aq)\,|\,Zn(s). \quad \textbf{(1 mark)}$$

Use the values for $E^\ominus$ given on this page.

Predicting reactions

$E^{\ominus}$ values can be used to predict if a given redox reaction is feasible.

The electrochemical series

Standard electrode potentials can be listed as an **electrochemical series**. Here is a part of it.

Right-hand electrode system	$E^{\ominus}$/V
$Mg^{2+}(aq) + 2e^- \rightleftharpoons Mg(s)$	-2.37
$Zn^{2+}(aq) + 2e^- \rightleftharpoons Zn(s)$	-0.76
$H^+(aq) + e^- \rightleftharpoons \frac{1}{2}H_2(g)$	0.00
$Cu^{2+}(aq) + 2e^- \rightleftharpoons Cu(s)$	$+0.34$
$Fe^{3+}(aq) + e^- \rightleftharpoons Fe^{2+}(aq)$	$+0.77$

oxidising agent — least powerful ↓ most powerful

reducing agent — most powerful ↑ least powerful

The Data Book given to you in the examination contains an electrochemical series, but it only shows the half-equations and $E^{\ominus}$ values (not the oxidising and reducing abilities of each species).

In this table:
- Iron(III) ions are the most powerful oxidising agent.
- Magnesium is the most powerful reducing agent.

Predicting the feasible reaction

An electrochemical reaction will be feasible if $E^{\ominus}_{cell}$ is positive. To predict the feasible reaction:

1 Write $E^{\ominus}$ for the two electrode systems side by side, most positive $E^{\ominus}$ on the right.

2 Reduction will be on the right, so write the half-equation as seen in the table.

3 Oxidation will be on the left, so write the other half-equation in reverse.

4 Combine the two half-equations, adjusting for uneven numbers of electrons if needed.

For example, is this reaction feasible?

$$MgSO_4(aq) + Zn(s) \rightarrow Mg(s) + ZnSO_4(aq)$$
$$-2.37\,V \qquad\qquad\qquad -0.76\,V$$
$$Mg \rightarrow Mg^{2+} + 2e^- \quad \rightarrow \quad Zn^{2+} + 2e^- \rightarrow Zn$$
$$\text{oxidation} \qquad e^- \text{ flow} \qquad \text{reduction}$$

Feasible reaction is:

$$Mg + Zn^{2+} \rightarrow Mg^{2+} + Zn$$

So no, the reaction at the top is not feasible.

Using the data table

A quick way to predict the feasible reaction is to use the data table directly: electrons will flow from the more negative electrode system towards the more positive electrode system.

Take care to choose the correct $E^{\ominus}$ values. It can be easy to select the wrong one for reactions involving copper or iron, for example.

Worked example

Use standard electrode potentials to explain why this disproportionation reaction happens:

$$2Cu^+(aq) \rightarrow Cu(s) + Cu^{2+}(aq) \qquad \textbf{(2 marks)}$$
1   $Cu^{2+}(aq) + e^- \rightleftharpoons Cu^+(aq) \quad E^{\ominus} = +0.15\,V$
2   $Cu^+(aq) + e^- \rightleftharpoons Cu(s) \quad E^{\ominus} = +0.52\,V$

The $E^{\ominus}$ value for equilibrium 2 is more positive than the $E^{\ominus}$ value for equilibrium 1. This means that the position of equilibrium 2 will move to the right and the position of equilibrium 1 will move to the left. As a result, Cu^+ ions can release electrons to form Cu^{2+} ions, and also accept electrons to form Cu atoms.

In terms of the method above:

1 Equilibrium 2 has the more positive $E^{\ominus}$
   Its half-equation is written on the right:
    $Cu^+(aq) + e^- \rightarrow Cu(s)$
2 The other half-equation is written in reverse:
    $Cu^+(aq) \rightarrow Cu^{2+}(aq) + e^-$
3 Combining the two half-equations gives you:
    $2Cu^+(aq) \rightarrow Cu(s) + Cu^{2+}(aq)$

$$E^{\ominus}_{cell} = E^{\ominus}_{right} - 0 = E^{\ominus}_{left}$$
$$= +0.52 - (+0.15)$$
$$= +0.37\,V$$

This is positive, so the reaction is feasible.

Now try this

Predict the outcome of the following. In each case, justify your answer and write an equation if a reaction happens. Use the data table on this page.
(a) Zinc powder is added to copper(II) sulfate solution.   **(2 marks)**
(b) Copper is added to dilute hydrochloric acid (a source of $H^+(aq)$ ions).   **(1 mark)**

Limitations of predictions

In practice, thermodynamically feasible reactions may not happen.

Thermodynamic feasibility

You can predict the thermodynamic feasibility of a chemical reaction using $E^\ominus$ values (page 96).

However, the reaction may not happen because:

 The activation energy, E_a, may be very large.

 The conditions for the reaction may not be standard.

In addition, a non-feasible reaction may happen when the conditions are changed from standard:

- The position of equilibrium of a half-cell reaction may change, changing its E.
- This changes the overall E_{cell}, which may become positive as a result.

Maths skills $E^\ominus_{cell}$ and $\Delta G^\ominus$

$$\Delta G^\ominus = -nFE^\ominus_{cell}$$

moles of electrons   Faraday constant

$\Delta G^\ominus = -T\Delta S^\ominus_{total}$ for a reaction at constant temperature, and nF is constant for a given cell.

Therefore, at a given temperature:

$E^\ominus_{cell} \propto \Delta S^\ominus_{total}$

replaces = and the constant

The standard emf of a cell is directly proportional to the total entropy change.

✓ Cell reaction is feasible if $\Delta S^\ominus_{total}$ is positive.

✓ $E^\ominus_{cell}$ is positive if $\Delta S^\ominus_{total}$ is positive.

✓ A cell reaction is feasible if $E^\ominus_{cell}$ is positive.

Changing the position of equilibrium

The position of equilibrium in a half-cell changes if the concentration of a reactant is changed.

For example $Mg^{2+}(aq) + 2e^- \rightleftharpoons Mg(s)$

$$E^\ominus = -2.37\,V$$

If the concentration of $Mg^{2+}(aq)$ is decreased:

- Position of equilibrium moves to the left.
- More electrons are released.
- The electrode becomes more negative.
- E becomes more negative.

For the cell, $Mg(s)\,|\,Mg^{2+}(aq) :: Cu^{2+}(aq)\,|\,Cu(s)$

$$E^\ominus_{cell} = +0.34 - (-2.37) = +2.71\,V$$

If $[Mg^{2+}(aq)]$ is decreased, E for the left hand half-cell becomes more negative, so E_{cell} increases.

Maths skills $E^\ominus_{cell}$ and K

$$\Delta G^\ominus = -nFE^\ominus_{cell}$$

$\Delta G^\ominus = -RT\ln K$ (see page 94 for more details)

Putting them together: $nFE^\ominus_{cell} = RT\ln K$

$$E^\ominus_{cell} = \frac{RT}{nF}\ln K$$

R, n and F are constants.

Therefore, at a given temperature, T: $E^\ominus_{cell} \propto \ln K$

The standard emf of a cell is directly proportional to the natural logarithm of the equilibrium constant.

✓ $E^\ominus_{cell}$ is positive if $K > 1$.

Worked example

In this cell, electrons flow through the external circuit from left to right. Explain why. **(3 marks)**

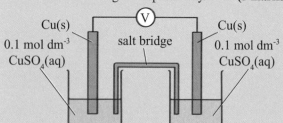

Cu(s)   salt bridge   Cu(s)
0.1 mol dm⁻³ CuSO₄(aq)   0.1 mol dm⁻³ CuSO₄(aq)

Copper(II) ions are more concentrated in the right hand half-cell. Copper(II) ions and electrons will be released in the left hand half-cell in preference to the right hand half-cell, so the left hand electrode is more negative.

The symbol Ⓥ represents a voltmeter. When a current flows, the measured potential difference decreases. When the current is zero, the reading is equal to $E^\ominus_{cell}$. The use of a high-resistance voltmeter allows negligible current to flow.

Now try this

State and explain the effect on the E_{cell} value for this cell if the concentration of the copper ion solution is decreased by adding water:
$Mg(s)\,|\,Mg^{2+}(aq) :: Cu^{2+}(aq)\,|\,Cu(s)$.   **(3 marks)**

Storage cells and fuel cells

Fuel cells generate a voltage by the reaction of hydrogen, methanol or other hydrogen-rich fuels with oxygen.

Hydrogen–oxygen fuel cells

Hydrogen reacts with oxygen to produce water:

$$H_2(g) + \tfrac{1}{2}O_2(g) \rightarrow H_2O(l) \quad \Delta H^\ominus = -286\,\text{kJ}\,\text{mol}^{-1}$$

Hydrogen can be used in internal combustion engines, similar to those used for petrol engines.

It is also used as the fuel for fuel cells. These:

- produce electricity directly from the reaction between hydrogen and oxygen (from air)
- produce electricity as long as hydrogen and oxygen are supplied
- make water vapour as their only product.

When in use, hydrogen–oxygen fuel cells:

✓ do not release carbon dioxide, a greenhouse gas

✗ need hydrogen, which is explosive and difficult to store.

Components of a fuel cell

The diagram shows the main components of a hydrogen–oxygen fuel cell with an acidic electrolyte (you do not need to know details of its construction). The electrodes are coated with platinum.

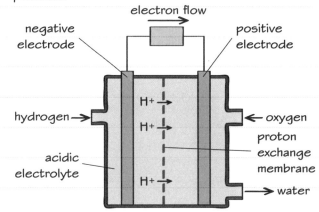

Fuel cells with an acidic electrolyte

The electrode reactions are:

- negative electrode:
$$H_2(g) \rightarrow 2H^+(aq) + 2e^-$$
- positive electrode:
$$\tfrac{1}{2}O_2(g) + 2H^+(aq) + 2e^- \rightarrow H_2O(l)$$

Hydrogen ions pass through a **proton exchange membrane** from the negative electrode to the positive electrode. Overall:

$$H_2(g) + \tfrac{1}{2}O_2(g) \rightarrow H_2O(l)$$

Fuel cells with an alkaline electrolyte

The electrode reactions are:

- negative electrode:
$$H_2(g) + 2OH^-(aq) \rightarrow 2H_2O(l) + 2e^-$$
- positive electrode:
$$\tfrac{1}{2}O_2(g) + H_2O(l) + 2e^- \rightarrow 2OH^-(aq)$$

Water diffuses through a membrane from the negative electrode to the positive electrode, and OH^- ions diffuse the other way. Overall:

$$H_2(g) + \tfrac{1}{2}O_2(g) \rightarrow H_2O(l)$$

Worked example

Lithium ion cells are rechargeable. The cell diagram for one type of the cell is shown below:

$$Li|Li^+::Li^+,MnO_2|LiMnO_2|Pt$$

(a) Write half-equations for the reactions at each electrode when the cell is being discharged (supplying a current). **(2 marks)**

Negative electrode: $Li \rightarrow Li^+ + e^-$

Positive electrode: $Li^+ + MnO_2 + e^- \rightarrow LiMnO_2$

(b) Write an equation for the overall reaction when the cell is being recharged. **(2 marks)**

$$\cancel{Li^+} + \cancel{e^-} + LiMnO_2 \rightarrow Li + \cancel{Li^+} + MnO_2 + \cancel{e^-}$$
$$LiMnO_2 \rightarrow Li + MnO_2$$

Rechargeable batteries are **storage cells**. They include lithium ion cells, nickel–cadmium cells and the lead–acid batteries used in vehicles.

Unlike non-rechargeable cells, they can be recharged because their cell reactions can be reversed by applying a potential difference.

Now try this

Nickel–cadmium cells ('nicads') are storage cells. During discharge, reactions are:

Negative electrode: $Cd + 2OH^- \rightarrow Cd(OH)_2 + 2e^-$

Positive electrode:

$$2NiO(OH) + 2H_2O + 2e^- \rightarrow 2Ni(OH)_2 + 2OH^-$$

Write an equation for the overall reaction when the cell is being recharged. **(2 marks)**

Exam skills 9

This exam-style question uses knowledge and skills you have already revised. Have a look at pages 95–99 for a reminder about **electrode potential** and **cells**.

Worked example

(a) (i) Describe the standard hydrogen electrode.

(4 marks)

Hydrogen gas at 100 kPa is bubbled through 1.0 mol dm^{-3} hydrochloric acid at 298 K in contact with a platinum electrode.

(ii) State why its standard electrode potential, $E^{\ominus}$, is 0.00 V. (1 mark)

It is defined as 0.00 V.

(b) (i) Use the Data Booklet to complete the table. (1 mark)

Half-equation	$E^{\ominus}$/V
$Zn^{2+} + 2e^- \rightleftharpoons Zn$	−0.76
$MnO_2 + H_2O + e^- \rightleftharpoons MnO(OH) + OH^-$	+0.74
$\frac{1}{2}O_2 + 2H^+ + 2e^- \rightleftharpoons H_2O$	+1.23

The diagram shows the structure of a 'heavy duty' zinc chloride non-rechargeable cell.

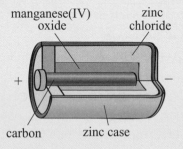

manganese(IV) oxide   zinc chloride

+   −

carbon   zinc case

(ii) Use values from the table completed in part (b) to calculate $E^{\ominus}_{cell}$ for this cell. (1 mark)

$E^{\ominus}_{cell} = E^{\ominus}_R - E^{\ominus}_L = +0.74 - (-0.76)$
$= +1.50\,V$

(iii) Write an equation for the overall reaction that occurs when the cell is in use. (1 mark)

$Zn + 2MnO_2 + 2H_2O$
$\rightarrow Zn^{2+} + 2MnO(OH) + 2OH^-$

(d) A methanol–oxygen fuel cell uses methanol as its fuel. When in use, methanol reacts with water to form carbon dioxide and hydrogen ions.

(i) Write a half-equation for the reaction that occurs at the methanol electrode. (1 mark)

$CH_3OH + H_2O \rightarrow CO_2 + 6H^+ + 6e^-$

(ii) The $E^{\ominus}_{cell}$ for a methanol–oxygen fuel cell is 1.21 V. Use data from the table above to calculate the $E^{\ominus}$ for the methanol electrode. (1 mark)

$E^{\ominus}_L = E^{\ominus}_R - E^{\ominus}_{cell} = +1.23 - 1.21 = +0.02\,V$

The substances needed are mentioned: $H_2(g)$ and HCl(aq). You could mention H^+ ions instead of a named strong acid.

The conditions needed are mentioned: do not confuse 298 K (25 °C) with 273 K (0 °C), or use incorrect units for pressure.

You cannot measure absolute electrode potentials. They are measured against the defined standard hydrogen electrode.

The values have been taken accurately from the Data Booklet. Take care to write the correct signs and numbers.

Remember that the Data Booklet given to you in the examination contains a table of standard electrode potentials.

The feasible cell reaction is the one where the half-cell with the more positive electrode potential is on the right.

In this question, that will be the manganese(IV) oxide half-cell, with the zinc half-cell on the left.

Remember that predictions made using $E^{\ominus}$ values are influenced by:
• high activation energies
• non-standard conditions.

Remember to multiply each half-equation as necessary so that both have the same number of electrons. Make sure you cancel species that appear on both sides.

You need to know the electrode reactions in a hydrogen–oxygen fuel cell.

Operating in acidic conditions:
$H_2 + \rightarrow 2H^+ + 2e^-$
$\frac{1}{2}O_2 + 2H^+ + 2e^- \rightarrow H_2O$

Operating in alkaline conditions:
$H_2 + 2OH^- \rightarrow 2H_2O + 2e^-$
$\frac{1}{2}O_2 + H_2O + 2e^- \rightarrow 2OH^-$

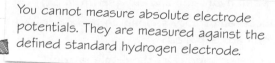

Reduction occurs on the right with the more positive electrode, so R must refer here to the $\frac{1}{2}O_2 + 2H^+ + 2e^- \rightleftharpoons H_2O$ electrode reaction (given in the table).

Redox titrations

Redox titrations are used to determine the concentration of a reactant such as iron(II) ions or iodine.

🧪 Practical skills — Manganate(VII) and iron(II)

Manganate(VII) ions can be used in redox titrations to determine the concentration of iron(II) ions.

The half-equations involved are:

$$MnO_4^-(aq) + 8H^+(aq) + 5e^- \rightarrow Mn^{2+}(aq) + 4H_2O(l)$$
$$Fe^{2+}(aq) \rightarrow Fe^{3+}(aq) + e^-$$

Combine them after multiplying the second equation by 5 (the first has 5 electrons):

purple $\quad MnO_4^-(aq) + 8H^+(aq) + 5Fe^{2+}(aq)$
$$\downarrow$$
pale pink $\quad Mn^{2+}(aq) + 4H_2O(l) + 5Fe^{3+}(aq)$

In a typical titration:

1. An iron(II) solution is acidified with dilute sulfuric acid (provide $H^+(aq)$).
2. $KMnO_4(aq)$ is added from a burette.
3. The $KMnO_4(aq)$ turns colourless in the iron(II) solution until the end point, then one more drop turns the mixture pale pink.

🧪 Practical skills — Thiosulfate and iodine

Thiosulfate ions can be used in redox titrations to determine the concentration of an iodine solution.

The half-equations involved are:

$$2S_2O_3^{2-}(aq) \rightarrow S_4O_6^{2-}(aq) + 2e^-$$
$$I_2(aq) + 2e^- \rightarrow 2I^-$$

You can combine them (both have 2 electrons):

$$2S_2O_3^{2-}(aq) + I_2(aq) \rightarrow S_4O_6^{2-}(aq) + 2I^-$$

In a typical titration:

1. An iodine solution is pipetted into a flask.
2. $Na_2S_2O_3(aq)$ is added from a burette.
3. The iodine solution turns from brown to pale yellow near the end point.
4. A few drops of starch indicator are added.
5. The mixture then turns from blue to colourless at the end point.

These titrations are often used to measure the concentration of iodine produced in a reaction.

Worked example

Iron(II) sulfate tablets are sold as a dietary supplement. One 0.5 g tablet was dissolved in dilute sulfuric acid. The titre was 22.40 cm³ when the solution was titrated against 0.010 mol dm⁻³ $KMnO_4(aq)$. Calculate the percentage of iron in the tablet. **(4 marks)**

Amount of $MnO_4^- = 0.010 \times 22.40 \times 10^{-3}$
$= 2.24 \times 10^{-4}$ mol

Amount of $Fe^{2+} = 5 \times 2.24 \times 10^{-4}$
$= 1.12 \times 10^{-3}$ mol

Mass of Fe $= 1.12 \times 10^{-3} \times 55.8$
$= 0.0625$ g

Percentage of iron $= 100 \times 0.0625 \div 0.5$
$= 12.5\%$

Worked example

A piece of brass was dissolved in excess nitric acid. The remaining acid was neutralised with sodium carbonate. A precipitate of copper(II) carbonate formed, which was then dissolved in dilute ethanoic acid. The mixture was made up to 250 cm³ with water and 25.0 cm³ of this was pipetted into a flask. Potassium iodide was added to liberate iodine, which was titrated against 0.200 mol dm⁻³ $Na_2S_2O_3(aq)$. The mean titre was 27.20 cm³. Calculate the mass of copper in the brass. **(6 marks)**

Amount of $S_2O_3^{2-} = 0.200 \times 27.20 \times 10^{-3}$
$= 5.44 \times 10^{-3}$ mol

Amount of $I_2 = 5.44 \times 10^{-3} \div 2$
$= 2.72 \times 10^{-3}$ mol

$$2Cu^{2+}(aq) + 4I^-(aq) \rightarrow 2CuI(s) + I_2(aq)$$

Amount of Cu^{2+} in 25.0 cm³
$= 2 \times 2.72 \times 10^{-3} = 5.44 \times 10^{-3}$ mol

Total mass of Cu in 250 cm³
$= 10 \times 5.44 \times 10^{-3} \times 63.5 = 3.45$ g

Now try this

A steel paper clip has a mass of 0.25 g. It was dissolved in acid and the iron oxidised (reduced) to Fe^{2+} using powdered zinc. The mixture was filtered and the filtrate made up to 250 cm³. 25.0 cm³ portions were titrated against 0.003 00 mol dm⁻³ $KMnO_4(aq)$. The mean titre was 26.50 cm³. Calculate the percentage of iron in the steel. **(5 marks)**

The overall ratios are $2S_2O_3^{2-} : I_2 : 2Cu^{2+}$
This means: amount of Cu^{2+} = amount of $S_2O_3^{2-}$
You multiply by 10 in the last step because a 25 cm³ sample was used from a total of 250 cm³.

d-block atoms and ions

Transition metals show variable oxidation numbers.

The d-block and transition metals

The d-block in the Periodic Table:

 occupies the area between groups 2 and 3

 contains the transition metals.

A **transition metal** is:

- a d-block element that
- forms one or more stable ions with incompletely filled d-orbitals.

The d-block and transition metals

You need to be able to work out the **electronic configurations** of atoms and ions of the d-block elements in Period 4 (Sc to Zn).

Using the definition opposite, scandium and zinc are <u>not</u> transition metals because their ions do not have incompletely filled d-orbitals:

- all transition metals are d-block elements
- but not all d-block elements are transition metals.

Electronic configurations of atoms

The full electronic configuration for scandium is:

$1s^2\ 2s^2\ 2p^6\ 3s^2\ 3p^6\ 3d^1\ 4s^2$

You will also see abbreviated configurations, where the $1s^2\ 2s^2\ 2p^6\ 3s^2\ 3p^6$ part is shown as [Ar] (argon is the noble gas in Period 3 that has this electronic configuration).

The 4s-orbital fills before 3d-orbitals but you may show 4s after 3d in electronic configurations.

You may need to draw **electrons in box diagrams**.

For example, for iron:

One 4s electron is promoted to a 3d orbital

3d   4s

[Ar] | ↑↓ | ↑ | ↑ | ↑ | ↑ | | ↑↓ |

Hund's rule: electrons occupy orbitals singly before pairing happens.

Pauli Exclusion Principle: two electrons cannot occupy the same orbital unless they have opposite spins (shown as ↑ and ↓ in the boxes).

Abbreviated electronic configurations for the d-block elements in period 4:

Z	Element		Electronic configuration
21	scandium	Sc	[Ar] $3d^1\ 4s^2$
22	titanium	Ti	[Ar] $3d^2\ 4s^2$
23	vanadium	V	[Ar] $3d^3\ 4s^2$
24	chromium	Cr	[Ar] $3d^5\ 4s^1$
25	manganese	Mn	[Ar] $3d^5\ 4s^2$
26	iron	Fe	[Ar] $3d^6\ 4s^2$
27	cobalt	Co	[Ar] $3d^7\ 4s^2$
28	nickel	Ni	[Ar] $3d^8\ 4s^2$
29	copper	Cu	[Ar] $3d^{10}\ 4s^1$
30	zinc	Zn	[Ar] $3d^{10}\ 4s^2$

Worked example

(a) Zinc only forms Zn^{2+} ions. Write the electronic configuration for this ion and explain why zinc is not a transition metal. **(3 marks)**

$1s^2\ 2s^2\ 2p^6\ 3s^2\ 3p^6\ 3d^{10}$

The Zn^{2+} ion has a completely filled d sub-shell. As the ion does not have an incompletely filled d-orbital, zinc cannot be a transition metal.

(b) Complete this diagram to show the electronic configuration for the Fe^{3+} ion. **(1 mark)**

3d   4s

[Ar] | ↑ | ↑ | ↑ | ↑ | ↑ | | |

Zinc commonly forms compounds in which zinc has oxidation number +2. It can form a small number of compounds in which its oxidation number is +1.

If you look at the electronic configuration for a Zn^+ ion, you can see that this does not have an incompletely filled d-orbital either:

$1s^2\ 2s^2\ 2p^6\ 3s^2\ 3p^6\ 3d^{10}\ 4s^1$

The 4s electrons are lost before any 3d electrons are lost when transition metal ions form. Take care if you do write electronic configurations with the 4s sub-shell before the 3d sub-shell.

Now try this

1 Write the electronic configuration for the V^{3+} ion. **(1 mark)**

2 Scandium only forms Sc^{3+} ions. Write the electronic configuration for this ion and explain why scandium is not a transition metal. **(3 marks)**

Ligands and complex ions

Transition metal ions can form complexes with ligands such as hydroxide ions and water.

Complex ions

Transition metals can form **complexes**.

These are species containing a central metal ion surrounded by ligands bonded to it by dative (co-ordinate) bonds.

Complex ions are complexes with an overall positive or negative charge.

For example, $[Cu(H_2O)_6]^{2+}$ and $[CuCl_4]^{2-}$.

Ligands and coordination number

A **ligand** is a species that can donate at least one pair of electrons to a metal ion, forming a dative bond. Ligands can be:
- molecules such as water or ammonia
- negative ions such as Cl^- or OH^-.

The **coordination number** is the number of dative bonds in a complex.

Be careful! Coordination number is <u>not</u> defined as the number of ligands in a complex.

Monodentate ligands

A **monodentate** ligand can donate one lone pair of electrons to the central metal ion.

Monodentate ligands include:

Ligand	Name in complexes
hydroxide ion $:OH^-$	hydroxo
ammonia $:NH_3$	ammine

Take care – not 'amine'

The chloride ion, Cl^- ('chloro'), and water, H_2O ('aqua'), are also monodentate ligands.

Although the oxygen atom in a water molecule has two lone pairs of electrons, only one of these can form a dative bond with the metal ion.

Bidentate ligands

A **bidentate** ligand can donate two lone pairs of electrons to the central metal ion.

1,2-diaminoethane, $NH_2CH_2CH_2NH_2$, has a lone pair of electrons on each nitrogen atom; each one can form a dative bond.

1,2-diaminoethane is often abbreviated to 'en' in equations involving complex ion. For example:

$$[Ni(H_2O)_6]^{2+} + 3en \rightarrow [Ni(en)_3]^{2+} + 6H_2O$$

Worked example

The $EDTA^{4-}$ ion can act as a ligand in complexes.
(a) The diagram shows the displayed formula for the $EDTA^{4-}$ ion. Draw the positions of the lone pairs of electrons available to form dative bonds with a metal ion. **(1 mark)**

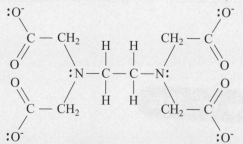

(b) Explain why $EDTA^{4-}$ may be described as a multidentate ligand. **(1 mark)**

It has several (six) lone pairs of electrons available to form dative bonds with a metal ion.

The $EDTA^{4-}$ ion is named after its non-systematic name, ethylenediaminetetraacetate.

Notice that the $EDTA^{4-}$ ion resembles 1,2-diaminoethane (above) but with a $-CH_2COO^-$ group replacing each hydrogen atom in the amino groups, $-NH_2$.

$EDTA^{4-}$ is also described as a hexadentate ligand but the A Level Specification only refers to the term 'multidentate'.

Now try this

The ethanedioate ion, $C_2O_4^{2-}$ can act as a ligand. With the help of a displayed formula, explain why the ethanedioate ion is a bidentate ligand. **(2 marks)**

The ethanedioate ion may also be represented as $^-OOC\text{–}COO^-$.

Shapes of complexes

Complex ions have different shapes, depending on the number of ligands around their metal ions.

Octahedral complexes

Complexes containing six ligands have **six-fold coordination**.

They have an **octahedral** shape with bond angles of 90°.

For example, hexaaquacopper(II), $[Cu(H_2O)_6]^{2+}$:

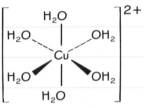

Remember, in 3D displayed formulae:

1   solid lines = bonds in the plane of the paper

2   dashed lines = bonds going behind the paper

3   wedges = bonds coming out of the paper

Complexes with this shape are usually formed when the ligands are H_2O, NH_3 and/or OH^-.

Tetrahedral complexes

Complexes containing four ligands have **four-fold coordination**.

They have a **tetrahedral** shape with bond angles of 109.5°.

For example, tetrachlorocuprate(II), $[CuCl_4]^{2-}$:

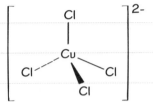

Complexes with this shape are usually formed when the ligands are relatively large, such as Cl^-.

Electron pair repulsion theory

Page 12 gives you detail about applying this theory to shapes of molecules and ions.

For complexes, you count the lone pairs of electrons donated to the central metal ion.

Square planar complexes

Complexes containing four ligands may also have a **square planar** shape, with 90° bond angles.

For example, $[PtCl_2(NH_3)_2]$:

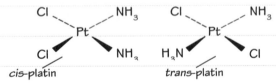

cis-platin   *trans*-platin

This complex has two **stereoisomers** (see page 45 for details about *cis–trans* and *E–Z* isomers in organic compounds):

- in *cis*-platin, the ammine ligands are next to each other,
- in *trans*-platin, the ammine ligands are opposite each other.

Naming complex ions

The systematic name for a complex ion tells you:

- the number and identity of each ligand
- the identity of the central metal ion and its oxidation number
- whether the complex ion is positively charged or negatively charged.

Different ligands in a complex are named in alphabetical order. For example:

$[Cu(H_2O)_5(OH)]^+$ is pentaaquahydroxocopper(II).

Metals have their usual name for neutral complexes or positively charged complex ions.

For negatively charged complex ions, their name (or their Latin name) ends with 'ate'. For example, $[FeCl_4]^{2-}$ is tetrachloroferrate(II).

Worked example

The complex, *cis*-platin, is used in cancer treatment. Explain why it is supplied as a single isomer and not in a mixture with the *trans* form. **(3 marks)**

The *cis* form has anti-cancer properties. The *trans* form is much less effective and is more toxic. Using just the single isomer reduces the risk of harmful side effects.

Full name: *cis*-diamminedichloroplatinum(II).

Now try this

Predict the formula, name and shape for the complex ions formed by:

(a) Chromium(III) ions and ammonia molecules. **(3 marks)**

(b) Manganese(II) ions and chloride ions. **(3 marks)**

You do not need to know how *cis*-platin works.

Colours

Transition metals form coloured ions in solution due to splitting of the energy levels of their d-orbitals by ligands.

The visible spectrum

Visible light is the part of the electromagnetic spectrum detected by the human eye.

The frequency of the light increases going from red to violet:

increasing frequency (energy) ⟶

When an atom or ion absorbs light:

- an electron is **promoted** from its normal **ground state** to a higher **excited** state.

A change from one energy level to another is called an **electron transition**.

The frequency of light absorbed depends upon the energy of the transition:

The greater the energy of the transition, the higher the frequency of the light absorbed.

Complementary colours

When white light passes through a solution, some frequencies of light may be absorbed.

The light that emerges comprises the remaining frequencies. For example, the solution in the diagram would appear blue-green because it absorbs red and orange light.

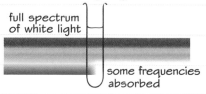

full spectrum of white light

some frequencies absorbed

The colours emerging from a solution are **complementary colours**, which are opposite each other on the 'colour wheel'.

purple
red   blue
orange   green
yellow

Splitting d-orbitals

When ligands bond with a transition metal ion, electrons in the ligands and electrons in the d-orbitals repel each other. As a result:

1 The energy levels of the d-orbitals increase.

2 Two slightly different energy levels form.

3 A d electron can be promoted from the lower level to the higher level, if this is partially occupied.

Transition metal ions have incompletely filled d-orbitals, so when light passes through a solution:

- Energy in the light promotes an electron from the lower level to the higher level.

Octahedral complexes

The d-orbitals in metal ions in octahedral complexes, such as Cu^{2+} in $[Cu(H_2O)_6]^{2+}$, split to produce two d-orbitals at a higher energy level.

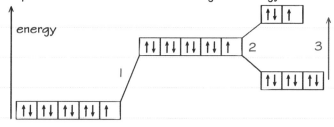

energy

Tetrahedral complexes

The d-orbitals in metal ions in tetrahedral complexes, such as Cu^{2+} in $[CuCl_4]^{2-}$, split to produce three d-orbitals at a higher energy level.

Worked example

(a) Explain why the hexaaquacopper(II) complex ion, $[Cu(H_2O)_6]^{2+}$, is blue. **(2 marks)**

The water ligands split the d-orbital energies, and d-d electron transitions absorb all light except for blue light.

(b) Explain why the hexaaquazinc(II) complex ion, $[Zn(H_2O)_6]^{2+}$, is colourless. **(2 marks)**

The d-orbitals of Zn^{2+} are full, so d-d electron transitions cannot happen.

Note: scandium(III) complexes are colourless because the d-orbitals of the Sc^{3+} ion are empty.

Now try this

Explain why the tetrachlorocuprate(II) complex ion, $[CuCl_4]^{2-}$, is green. **(2 marks)**

Colour changes

Colour changes in transition metal ions may occur because of a change in oxidation number, ligand or coordination number.

Change in oxidation number

For a given transition metal ion, the higher its **oxidation number**:

- the greater the amount of splitting of the energy levels of the d-orbitals by ligands
- the greater the energy level difference becomes.

This affects the particular frequencies of light absorbed by a complex ion, changing its colour.

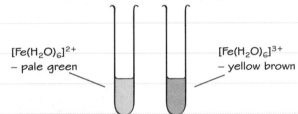

$[Fe(H_2O)_6]^{2+}$
– pale green

$[Fe(H_2O)_6]^{3+}$
– yellow brown

Change in ligand

Different **ligands** cause different amounts of splitting of the energy levels of the d-orbitals:

Ligand	Splitting	Energy level difference
Cl⁻	least	smallest
OH⁻	↓	↓
H₂O		
NH₃	most	greatest

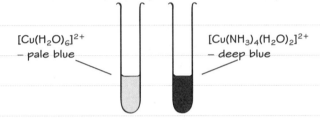

$[Cu(H_2O)_6]^{2+}$
– pale blue

$[Cu(NH_3)_4(H_2O)_2]^{2+}$
– deep blue

Haemoglobin

Haemoglobin is the protein found in red blood cells that carries oxygen in the blood. It is an iron(II) complex containing a multidentate ligand.

It binds reversibly with oxygen:

haemoglobin + oxygen ⇌ oxyhaemoglobin
  purple-blue   bright red

Carbon monoxide, CO, is a product of incomplete combustion (page 47 has details about this and the toxicity of carbon monoxide).

The carbon atom has a lone pair of electrons:

 Carbon monoxide can act as a ligand.

2 It binds to haemoglobin more strongly than oxygen does.

A **ligand exchange** reaction happens:
(bright red) oxyhaemoglobin + carbon
                                              ↓ monoxide
(cherry red) carboxyhaemoglobin + oxygen

Change in coordination number

Coordination number affects an ion's colour:

Coordination number	4		6
Shape	tetrahedral		octahedral
Splitting	least ⟶		most
Energy level difference	least ⟶		most

However, a change in coordination number also involves a change in ligand, so any difference in colour is due to a combination of <u>both</u> factors.

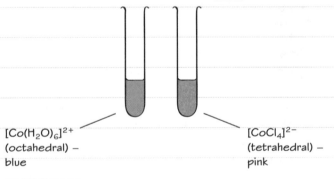

$[Co(H_2O)_6]^{2+}$
(octahedral) –
blue

$[CoCl_4]^{2-}$
(tetrahedral) –
pink

The two complex ions have the same coordination number (6) and ligand (water or aqua).

The energy level difference is higher in the $[Cr(H_2O)_6]^{3+}$ ion, so it can absorb higher frequency light than the $[Cr(H_2O)_6]^{2+}$ ion can.

Worked example

Suggest why the $[Cr(H_2O)_6]^{2+}$ ion is sky blue but the $[Cr(H_2O)_6]^{3+}$ ion is violet.   **(2 marks)**

Their central metal ions have different oxidation numbers (+2 and +3), so the energy levels of their d-orbitals are split by different amounts.

Now try this

Suggest why carboxyhaemoglobin and oxyhaemoglobin are different colours.   **(2 marks)**

Had a look ☐ Nearly there ☐ Nailed it! ☐

Vanadium chemistry

Vanadium can be reduced from oxidation number +5 to +2 by zinc in acidic solution.

Colours of vanadium compounds and the oxidation number of vanadium

The table summarises the colours of solutions containing vanadium ions.

Oxidation number	+5	+4	+3	+2
Formula	VO_2^+	VO^{2+}	V^{3+}	V^{2+}
Name	dioxovanadium(V)	oxovanadium(IV)	vanadium(III)	vanadium(II)
Colour of solution	yellow	blue	green	purple

Be careful! You need to know these colours. Take care not to confuse VO_2^+ with VO^{2+}.

Reduction from V(V) to V(II)

Ammonium trioxovanadate(V), NH_4VO_3, is a soluble vanadium(V) compound.

In acidic conditions, it forms the dioxovanadium(V) ion, VO_2^+.

This can be reduced to vanadium(II) using:
- zinc with sulfuric or hydrochloric acid.

You see a change in colour during the reaction:
1. yellow to blue (+3 to +4)
2. blue to green (+4 to +5)
3. green to purple (+3 to +2)

Explaining reduction using $E^{\ominus}$ values

The table summarises, in terms of standard electrode potentials, why these reactions happen.

Change	Oxidation (left)	Reduction (right)	$E^{\ominus}_{cell} = E^{\ominus}_{right} - E^{\ominus}_{left}$	Overall
+5 to +4	$Zn \rightleftharpoons Zn^{2+} + 2e^-$	$VO_2^+ + 2H^+ + e^- \rightleftharpoons VO^{2+} + H_2O$	$+1.00 - (-0.76)$ $= +1.76\,V$	$2VO_2^+ + 4H^+ + Zn$ $\downarrow$ $2VO^{2+} + 2H_2O + Zn^{2+}$
+4 to +3	$Zn \rightleftharpoons Zn^{2+} + 2e^-$	$VO^{2+} + 2H^+ + e^- \rightleftharpoons V^{3+} + H_2O$	$+0.34 - (-0.76)$ $= +1.10\,V$	$2VO^{2+} + 4H^+ + Zn$ $\downarrow$ $2V^{3+} + 2H_2O + Zn^{2+}$
+ 3 to +2	$Zn \rightleftharpoons Zn^{2+} + 2e^-$	$V^{3+} + e^- \rightleftharpoons V^{2+}$	$-0.26 - (-0.76)$ $= +0.50\,V$	$2V^{3+} + Zn$ $\downarrow$ $2V^{2+} + Zn^{2+}$

Notice that all three $E^{\ominus}_{cell}$ values are positive, so the reactions are feasible.

Worked example

Explain why V^{2+} ions cannot be reduced to vanadium using acidified zinc. **(2 marks)**

$E^{\ominus}_{cell} = E^{\ominus}_{right} - E^{\ominus}_{left} = -1.18 - (-0.76)$
$= -0.42\,V$

As $E^{\ominus}_{cell}$ is negative, the reaction is not feasible.

The Data Book shows you that, for the reaction:
$V^{2+} + 2e^- \rightleftharpoons V, E^{\ominus} = -1.18\,V.$

The overall reaction required would be:
$V^{2+} + Zn \rightarrow V + Zn^{2+}$

Since this is not feasible, the reduction reactions described above do not continue to vanadium.

Now try this

An excess of acidified potassium manganate(VII) solution, $KMnO_4(aq)/H^+(aq)$ was added to a solution containing $V^{2+}(aq)$ ions. Identify the vanadium species present when the reaction is complete and write the half-equation for its formation. **(2 marks)**

$MnO_4^- + 8H^+ + 5e^- \rightleftharpoons Mn^{2+} + 4H_2O, E^{\ominus} = +1.51\,V$

Right hand electrode system	$E^{\ominus}/V$
$Zn^{2+} + 2e^- \rightleftharpoons Zn$	$^-0.76$
$V^{3+} + e^- \rightleftharpoons V^{2+}$	$^-0.26$
$VO^{2+} + 2H^+ + e^- \rightleftharpoons V^{3+} + H_2O$	$+0.34$
$VO_2^+ + 2H^+ + e^- \rightleftharpoons VO^{2+} + H_2O$	$+1.00$

Chromium chemistry

Chromium can be reduced from oxidation number +6 to +2 by zinc in acidic solution.

Colours of chromium compounds and the oxidation number of chromium

The table summarises the colours of solutions containing chromium ions.

Oxidation number	+6	+6	+3	+2
Formula	CrO_4^{2-}	$Cr_2O_7^{2-}$	Cr^{3+}	Cr^{2+}
Name	chromate(VI)	dichromate(VI)	chromium(III)	chromium(II)
Colour of solution	yellow	orange	green	blue

Reduction from Cr(VI) to Cr(II)

Potassium dichromate(VI), $K_2Cr_2O_7$, is a soluble chromium(VI) compound. Its $Cr_2O_7^{2-}$ ions can be reduced to chromium(II) using:

- zinc with sulfuric or hydrochloric acid.

You see a change in colour during the reaction:

- orange to green (+6 to +3)
- green to blue (+3 to +2).

Standard electrode potentials, $E^\ominus$

Right hand electrode system	$E^\ominus$/V	e^- flow
$Zn^{2+} + 2e^- \rightleftharpoons Zn$	−0.76	↓
$Cr^{3+} + e^- \rightleftharpoons Cr^{2+}$	−0.41	
$Cr^{3+} + 3\frac{1}{2}H_2O \rightleftharpoons \frac{1}{2}Cr_2O_7^{2-} + 7H^+ + 3e^-$	+1.33	

The third equation is more usually written as:

$$Cr_2O_7^{2-} + 14H^+ + 6e^- \rightleftharpoons 2Cr^{3+} + 7H_2O$$

Explaining reduction using $E^\ominus$ values

The table summarises, in terms of standard electrode potentials, why these reactions happen.

Change	Oxidation (left)	Reduction (right)	$E^\ominus_{cell} = E^\ominus_{right} - E^\ominus_{left}$	Overall
+6 to +3	$Zn \rightleftharpoons Zn^{2+} + 2e^-$	$Cr_2O_7^{2-} + 14H^+ + 6e^- \rightleftharpoons 2Cr^{3+} + 7H_2O$	+1.33 − (−0.76) = +2.09 V	$Cr_2O_7^{2-} + 14H^+ + 3Zn$ ↓ $2Cr^{3+} + 7H_2O + 3Zn^{2+}$
+3 to +2	$Zn \rightleftharpoons Zn^{2+} + 2e^-$	$Cr^{3+} + e^- \rightleftharpoons Cr^{2+}$	−0.41 − (−0.76) = +0.35 V	$2Cr^{3+} + Zn$ ↓ $2Cr^{2+} + Zn^{2+}$

Notice that all three $E^\ominus_{cell}$ values are positive, so the reactions are feasible.

Worked example

$Cr_2O_7^{2-}$ ions can be converted into CrO_4^{2-} ions as a result of this equilibrium:

$$2CrO_4^{2-}(aq) + 2H^+(aq) \rightleftharpoons Cr_2O_7^{2-}(aq) + H_2O(l)$$

Describe the colour seen in acidic conditions, and in alkaline conditions. **(2 marks)**

The solution is orange in acidic conditions, due to $Cr_2O_7^{2-}(aq)$ ions, and yellow in alkaline conditions, due to $CrO_4^{2-}(aq)$ ions.

Note that the oxidation number of chromium does not change in this reaction – it is +6 in both ions.

Now try this

(a) Write an equation to represent the reduction of $Cr_2O_7^{2-}$ ions to Cr^{2+} ions by reaction with a mixture of zinc and sulfuric acid. **(2 marks)**

(b) If air is allowed back into the mixture containing Cr^{2+} ions, the colour changes from blue to green. Explain this observation. **(2 marks)**

Reactions with hydroxide ions

Solutions of transition metal ions undergo acid–base reactions with hydroxide ions to form precipitates.

Reactions of M^{2+}(aq) with OH^- ions

Solutions of M^{2+} transition metal ions react with hydroxide ions to produce coloured precipitates:

$$[M(H_2O)_6]^{2+}(aq) + 2OH^-(aq) \rightarrow [M(H_2O)_4(OH)_2](s) + 2H_2O(l)$$

pale blue solution   blue precipitate
$$[Cu(H_2O)_6]^{2+}(aq) + 2OH^-(aq) \rightarrow [Cu(H_2O)_4(OH)_2](s) + 2H_2O(l)$$

Gradually oxidises in air to brown $[Fe(H_2O)_3(OH)_3](s)$.

pale green solution   green precipitate
$$[Fe(H_2O)_6]^{2+}(aq) + 2OH^-(aq) \rightarrow [Fe(H_2O)_4(OH)_2](s) + 2H_2O(l)$$

Gradually turns pink on standing.

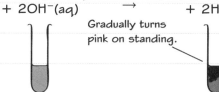

pale pink solution   blue precipitate
$$[Co(H_2O)_6]^{2+}(aq) + 2OH^-(aq) \rightarrow [Co(H_2O)_4(OH)_2](s) + 2H_2O(l)$$

Reactions of M^{3+}(aq) with OH^- ions

Solutions of M^{3+} transition metal ions react with hydroxide ions to produce coloured precipitates:

$$[M(H_2O)_6]^{3+}(aq) + 3OH^-(aq) \rightarrow [M(H_2O)_3(OH)_3](s) + 3H_2O(l)$$

These reactions are **acid–base** reactions (not ligand exchange – see page 110):

- An O–H bond in a water (aqua) ligand breaks, releasing an H^+ ion.
- An OH^- ion accepts the H^+ ion (it acts as a **Brønsted–Lowry base**, a proton acceptor).

Two water molecules are involved in the acid–base reactions of M^{2+} ions, and three for M^{3+} ions.

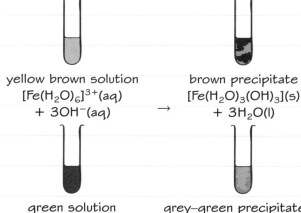

yellow brown solution   brown precipitate
$$[Fe(H_2O)_6]^{3+}(aq) + 3OH^-(aq) \rightarrow [Fe(H_2O)_3(OH)_3](s) + 3H_2O(l)$$

green solution   grey–green precipitate
$$[Cr(H_2O)_6]^{3+}(aq) + 3OH^-(aq) \rightarrow [Cr(H_2O)_3(OH)_3](s) + 3H_2O(l)$$

The $[Cr(H_2O)_6]^{3+}$ ion is violet, but you usually see green complexes (due to Cl^- and other ligands).

Worked example

A student added sodium hydroxide solution to chromium(III) sulfate solution. A grey-green precipitate of $[Cr(H_2O)_3(OH)_3]$(s) formed, which redissolved to form a dark green solution when excess sodium hydroxide solution was added.

(a) Write an equation for the reaction when excess sodium hydroxide solution was added. **(1 mark)**

$$[Cr(H_2O)_3(OH)_3](s) + 3OH^-(aq) \rightarrow [Cr(OH)_6]^{3-}(aq) + 3H_2O(l)$$

The student added hydrogen peroxide solution to the dark green solution, then warmed the mixture.

(b) State the colour of the resulting solution and write an equation for the reaction. **(2 marks)**

$$2[Cr(OH)_6]^{3-}(aq) + 3H_2O_2(aq) \rightarrow 2CrO_4^{2-}(aq) + 2OH^-(aq) + 8H_2O(l)$$

Chromium is oxidised from +3 to +6 in this reaction (hydrogen peroxide is acting as an **oxidising agent**). If acid is then added, the $Cr_2O_7^{2-}$ ion forms (see page 108 for details of the reaction).

Now try this

Aluminium is not a transition metal. However, its $[Al(H_2O)_6]^{3+}$(aq) ions react with hydroxide ions to produce a white precipitate of $[Al(H_2O)_3(OH)_3]$(s), which redissolves in excess alkali to form colourless $[Al(H_2O)_2(OH)_4]^-$(aq). Write equations to show these reactions. **(2 marks)**

Reactions with ammonia

Solutions of transition metal ions undergo acid–base reactions with ammonia to form precipitates.

Reactions of M^{2+}(aq) ions with NH_3

Solutions of M^{2+} transition metal ions react with ammonia to produce coloured precipitates:

$$[M(H_2O)_6]^{2+}(aq) \quad [M(H_2O)_4(OH)_2](s)$$
$$+\ 2NH_3(aq) \rightarrow +\ 2NH_4^+(aq)$$

These reactions are **acid–base** reactions:
- NH_3 acts as a Brønsted–Lowry base, but
- the precipitates are identical to those formed by reaction with OH^- ions (page 109).

Copper(II) and cobalt(II) hydroxide both redissolve in excess ammonia solution in **ligand exchange** reactions – iron(II) hydroxide does not.

blue precipitate
$[Co(H_2O)_4(OH)_2](s)$
$+\ 6NH_3(aq)$ →

brown solution
$[Co(NH_3)_6]^{2+}(aq)$
$+\ 4H_2O(l) + 2OH^-(aq)$

All six ligands (H_2O and OH^-) are replaced by NH_3. The brown solution darkens on standing, due to oxidation to $[Co(NH_3)_6]^{3+}(aq)$ by oxygen in the air.

blue precipitate
$[Cu(H_2O)_4(OH)_2](s)$
$+\ 4NH_3(aq)$ →

deep blue solution
$[Cu(NH_3)_4(H_2O)_2]^{2+}(aq)$
$+\ 2H_2O(l) + 2OH^-(aq)$

Be careful! Both OH^- ligands, but only two of the H_2O ligands, are replaced by NH_3 in copper(II) hydroxide.

Reactions of M^{3+}(aq) ions with NH_3

Solutions of M^{3+} transition metal ions react with ammonia to produce coloured precipitates:

$$[M(H_2O)_6]^{3+}(aq) \quad [M(H_2O)_3(OH)_3](s)$$
$$+\ 3NH_3(aq) \rightarrow +\ 3NH_4^+(aq)$$

These reactions are **acid–base** reactions:
- NH_3 acts as a Brønsted–Lowry base, but
- the precipitates are identical to those formed by reaction with OH^- ions (page 109).

Chromium(III) hydroxide redissolves in excess ammonia solution in a **ligand exchange** reaction – iron(III) hydroxide does not.

grey–green precipitate
$[Cr(H_2O)_3(OH)_3](s)$
$+\ 6NH_3(aq)$ →

purple solution
$[Cr(NH_3)_6]^{3+}(aq)$
$+\ 3H_2O(l) + 3OH^-(aq)$

All six ligands (H_2O and OH^-) are replaced by NH_3.

Amphoteric hydroxides

Chromium(III) hydroxide is **amphoteric**:

✓ It as acts as both an acid and as a base.

Acting as an acid (reacting with excess alkali):
$$3OH^-(aq) + \rightleftharpoons 3H_2O(l) +$$
$$[Cr(H_2O)_3(OH)_3]^{3+}(s) \quad [Cr(OH)_6]^{3-}(aq)$$
grey–green precipitate   dark green solution

Acting as a base (reacting with excess acid):
$$3H_3O^+(aq) + \rightleftharpoons 3H_2O(l) +$$
$$[Cr(H_2O)_3(OH)_3]^{3+}(s) \quad [Cr(H_2O)_6]^{3+}(aq)$$
grey-green precipitate   green solution

Worked example

Precipitates form when dilute ammonia solution is added to iron(II) sulfate solution or to iron(III) sulfate solution. State the colour of each precipitate, and write equations for their formation. **(4 marks)**

The iron(II) hydroxide precipitate is green:
$$[Fe(H_2O)_6]^{2+} + 2NH_3 \rightarrow [Fe(H_2O)_4(OH)_2] + 2NH_4^+$$

The iron(III) hydroxide precipitate is brown:
$$[Fe(H_2O)_6]^{3+} + 3NH_3 \rightarrow [Fe(H_2O)_3(OH)_3] + 3NH_4^+$$

These are acid–base reactions in which:
- an O–H bond in a water ligand breaks, releasing an H^+ ion
- an NH_3 molecule accepts the H^+ ion (it acts as a Brønsted–Lowry base, a proton acceptor).

Two water ligands are involved when iron(II) hydroxide forms, and three when iron(III) hydroxide forms.

Ligand exchange does not happen – neither precipitate redissolves in excess ammonia solution.

Now try this

Aluminium hydroxide is amphoteric.

Aluminium is not a transition metal. However, a white precipitate of $[Al(H_2O)_3(OH)_3](s)$ redissolves in excess acid to form a colourless solution containing $[Al(H_2O)_6]^{3+}(aq)$ ions. Write an equation to show this reaction. **(1 mark)**

Ligand exchange

Ligand exchange is accompanied by a colour change and may lead to a change in coordination number.

Copper(II) octahedral complexes

These copper(II) complexes all have coordination number 6 and an octahedral shape:

hexaaquacopper(II)

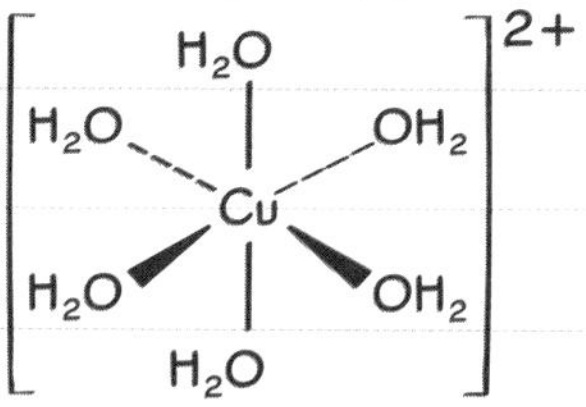

tetraaquadihydroxocopper(II)

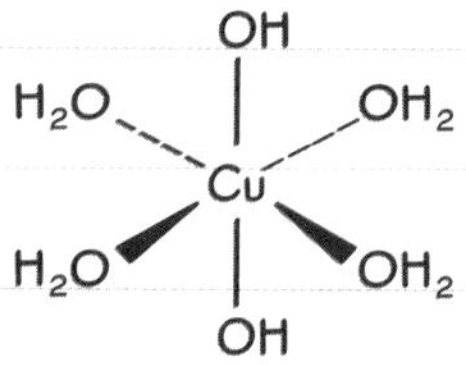

tetraamminediaquacopper(II)

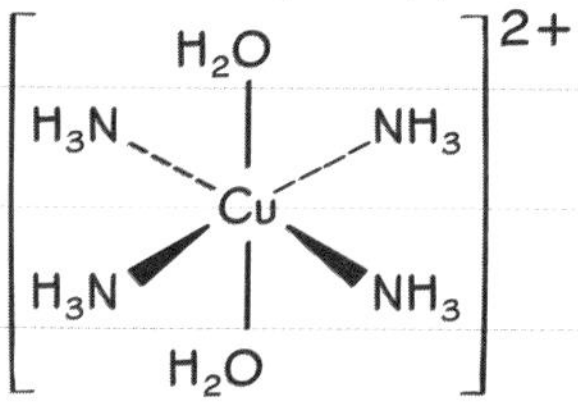

$[Cu(H_2O)_6]^{2+}$
pale blue

add NaOH(aq)
or NH_3(aq)
→

$[Cu(H_2O)_4(OH)_2]$
blue

add excess
NH_3(aq)
→

$[Cu(NH_3)_4(H_2O)_2]^{2+}$
deep blue

There is no change of oxidation number or coordination number, so the colour changes are due to the presence of different ligands (see page 106 for more detail about this).

Tetrahedral complexes

A change in coordination number can happen if:

- small, uncharged ligands such as H_2O are replaced by larger, charged ligands such as Cl^- ions.

This change happens to these complexes, (the coordination number changes from 6 to 4):

hexaaquacopper(II)   tetrachlorocuprate(II) 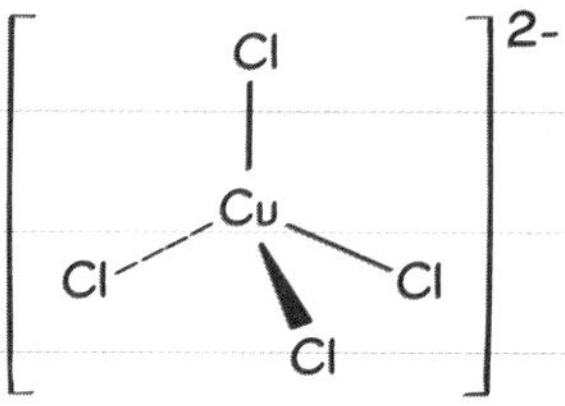   hexaaquacobalt(II) 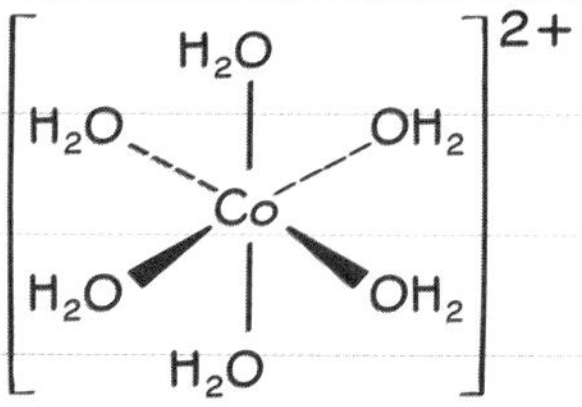   tetrachlorocobaltate(II)

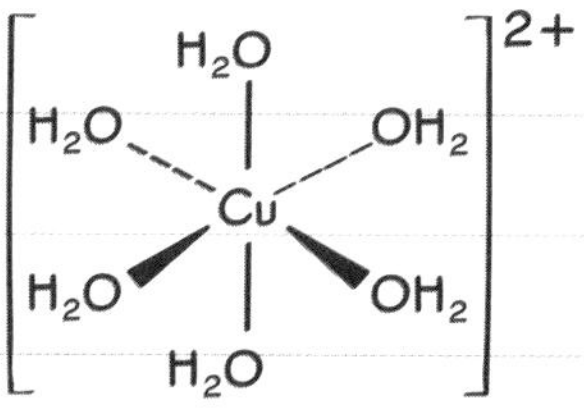

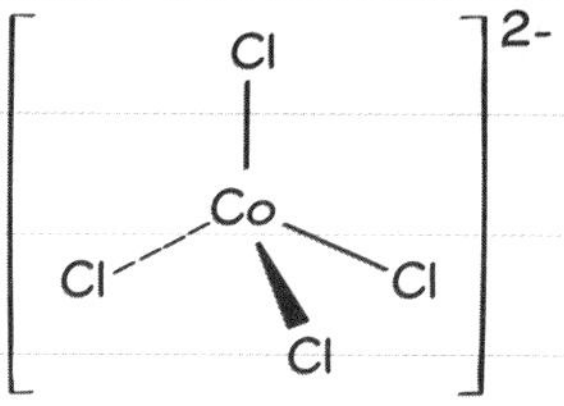

and

$[Cu(H_2O)_6]^{2+}$
pale blue

$[CuCl_4]^{2-}$
yellow

and

$[Co(H_2O)_6]^{2+}$
pink

$[CoCl_4]^{2-}$
blue

$[Cu(H_2O)_6]^{2+} + 4Cl^- \rightleftharpoons [CuCl_4]^{2-} + 6H_2O$
add conc. HCl(aq)   →

$[Co(H_2O)_6]^{2+} + 4Cl^- \rightleftharpoons [CoCl_4]^{2-} + 6H_2O$
add conc. HCl(aq) →

There is no change on oxidation number, but the coordination number and ligands change.

Worked example

A complex ion forms when 1,2-diaminoethane is added to an aqueous solution of copper(II) ions:

$[Cu(H_2O_6]^{2+} + 3H_2NCH_2CH_2NH_2$
$\rightarrow [Cu(H_2NCH_2CH_2NH_2)_3]^{2+} + 6H_2O$

Suggest, in terms of entropy change, why the complex ion formed is more stable than the $[Cu(H_2O)_6]^{2+}$ ion. **(2 marks)**

There is an increase in the number of particles in the reaction, from four to seven. This means that the entropy change of the system is positive.

1,2-diaminoethane is a bidentate ligand.
It is often represented as en in equations.
For example:

$[Cu(H_2O_6]^{2+} + 3en \rightarrow [Cu(en)_3]^{2+} + 6H_2O$

It is important that the answer refers to the entropy change of the system, ΔS_{system}, not just to entropy in general.

Now try this

A complex ion forms when $EDTA^{4-}$ is added to an aqueous solution of copper(II) ions:

$[Cu(H_2O_6]^{2+} + EDTA^{4-} \rightarrow [Cu(EDTA)]^{2-} + 6H_2O$

Suggest why the complex ion is more stable than the $[Cu(H_2O)_6]^{2+}$ ion.

(2 marks)

Heterogeneous catalysis

Heterogeneous catalysts are usually solids that catalyse reactions involving liquids and/or gases.

Heterogeneous catalysts

A **heterogeneous catalyst** is a catalyst that is in a different phase from the reactants.

The table shows examples from the Specification.

Catalyst	Used in
Iron	the Haber process
Vanadium(V) oxide, V_2O_5	the contact process
Nickel	addition reactions of alkenes with hydrogen
Platinum	catalytic converters

The reaction happens at the surface of the catalyst.

Change in oxidation number in V_2O_5

Sulfuric acid is manufactured by the **contact process**. The second stage is catalysed by V_2O_5:

$$SO_2(g) + \tfrac{1}{2}O_2(g) \rightleftharpoons SO_3(g)$$

Two redox reactions are involved:

1 $V_2O_5(s) + SO_2(g) \rightarrow V_2O_4(s) + SO_3(g)$
Vanadium is reduced from +5 to +4

2 $V_2O_4(s) + \tfrac{1}{2}O_2(g) \rightarrow V_2O_5(s)$
Vanadium is oxidised from +4 to +5

Overall, the oxidation state of vanadium is unchanged (and SO_2 is oxidised by O_2 to SO_3).

Adsorption, reaction, desorption

There are three main steps in the action of a heterogeneous catalyst:

1   **adsorption** of one or more reactants on to the surface of the catalyst

2   weakening of bonds in the reactants followed by reaction

3   **desorption** of products from the surface.

For example, manganese(IV) oxide catalyses the decomposition of hydrogen peroxide:

$$2H_2O_2(aq) \rightarrow 2H_2O(l) + O_2(g)$$

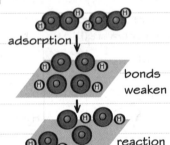

adsorption

bonds weaken

reaction

desorption

Poisoning

Catalysts can become 'poisoned' by impurities that adsorb strongly on to their **active sites** on the surface, blocking the adsorption of reactants. For example, lead poisons vehicle catalytic converters, reducing their efficiency.

Catalytic converters decrease the emissions of carbon monoxide and nitrogen monoxide from internal combustion engines.

(a) Write an equation to show how these two gases may produce less harmful products. **(1 mark)**

$CO(g) + NO(g) \rightarrow CO_2(g) + \tfrac{1}{2}N_2(g)$

(b) Explain how this reaction is catalysed. **(3 marks)**

CO and NO molecules are adsorbed onto the surface of the catalyst. Bonds in the molecules weaken and the chemical reaction occurs. The CO_2 and N_2 product molecules are then desorbed from the surface of the catalyst.

Do not muddle up the two words 'adsorption' and 'absorption' — they mean two different things.

Carbon monoxide is a toxic gas that binds to haemoglobin (see page 106 for more details). Nitrogen monoxide is one of the gases involved in causing acid rain (page 47 has more about this).

Nitrogen is the main component of air (78%). Carbon dioxide is a **greenhouse gas** implicated in global warming and climate change.

Silver forms very weak bonds with reactant molecules and tungsten forms very strong bonds with product molecules. Suggest why these two transition metals are poor catalysts. **(2 marks)**

Homogeneous catalysis

Homogeneous catalysts are usually in solution and catalyse reactions involving solutions.

Homogeneous catalysts

A **homogeneous catalyst** is a catalyst that is in the same phase as the reactants.

The table shows examples from the Specification.

Acts as an **autocatalyst**

Catalyst	Catalyses reaction between
Fe^{2+}	$I^-(aq)$ and $S_2O_8^{2-}(aq)$
Mn^{2+}	$MnO_4^-(aq)$ and $C_2O_4^{2-}(aq)$

Homogeneous catalysis proceeds via an **intermediate species**.

$S_2O_8^{2-} + I^-$ catalysed by Fe^{2+}

There are two steps in this reaction:

1 Reaction between Fe^{2+} and $S_2O_8^{2-}$

$$2Fe^{2+}(aq) + S_2O_8^{2-}(aq)$$
$$\rightarrow 2SO_4^{2-}(aq) + 2Fe^{3+}(aq)$$

Iron is oxidised from +2 to +3

2 Reaction between Fe^{3+} and I^-

$$2Fe^{3+} + 2I^-(aq) \rightarrow I_2(aq) + 2Fe^{2+}(aq)$$

Iron is reduced from +3 to +2

Overall, the reaction is:

$$S_2O_8^{2-}(aq) + 2I^-(aq) \rightarrow 2SO_4^{2-}(aq) + I_2(aq)$$

Note that Fe^{3+} could also catalyse this reaction:
- Fe^{2+} changes to Fe^{3+}, and back again
- the same two steps are involved, but in the reverse order

Charged reactant species

In the two reactions opposite, both reactant species are negatively charged.

They repel each other, leading to:
- a high activation energy, E_a, and
- a low rate of reaction at room temperature.

The catalysts for the reactions are positively charged. They will attract the reactant species.

🖩 Maths skills   Feasibility

You can use standard electrode potentials, $E^\ominus$, to show that the reactions opposite are all feasible under standard conditions.

Right-hand electrode system	$E^\ominus$/V
$\frac{1}{2}I_2 + e^- \rightleftharpoons I^-$	+0.54
$Fe^{3+} + e^- \rightleftharpoons Fe^{2+}$	+0.77
$\frac{1}{2}S_2O_8^{2-} + e^- \rightleftharpoons SO_4^{2-}$	+2.01

Remember: $E^\ominus_{cell} = E^\ominus_{right} - E^\ominus_{left}$

Reaction	$E^\ominus_{cell}$/V
Step 1	$+2.01 - (+0.77) = +1.24$
Step 2	$+0.77 - (+0.54) = +0.23$
Overall	$+2.01 - (+0.54) = +1.47$

A reaction will be feasible if $E^\ominus_{cell}$ is positive.

Worked example

At a constant temperature, the reaction between $C_2O_4^{2-}$ ions and MnO_4^- ions in acidic solution happens slowly to begin with, then becomes faster.

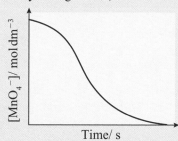

(a) Write an equation for the reaction between MnO_4^- ions and $C_2O_4^{2-}$ ions. **(1 mark)**

$2MnO_4^- + 5C_2O_4^{2-} + 16H^+ \rightarrow 2Mn^{2+} + 10CO_2 + 8H_2O$

(b) Mn^{2+} ions act as an autocatalyst for this reaction. State the meaning of the term 'autocatalyst', and write two equations to show how Mn^{2+} ions act as an autocatalyst. **(3 marks)**

An autocatalyst is a product of the reaction that acts as a catalyst for that reaction.
Step 1: $MnO_4^- + 4Mn^{2+} + 8H^+ \rightarrow 4H_2O + 5Mn^{3+}$
Step 2: $C_2O_4^{2-} + 2Mn^{3+} \rightarrow 2CO_2 + 2Mn^{2+}$

> Notice how manganese is oxidised from +2 to +3 in Step 1, then reduced from +3 to +2 again in Step 2.

> Both reactant species are negatively charged. They will repel each other in solution leading to a high activation energy.

Now try this

1 Explain why sodium ions, Na^+, cannot act as a homogeneous catalyst. **(1 mark)**

2 State why Fe^{2+} ions do not act as an autocatalyst in the reaction between $S_2O_8^{2-}$ ions and I^- ions. **(1 mark)**

Exam skills 10

This exam-style question uses knowledge and skills you have already revised. Have a look at pages 102–106 and 111–113 for a reminder about **transition metal chemistry** and **catalysis**.

Worked example

(a) Give the full electronic configurations for:

   (i) a chromium atom. **(1 mark)**

$1s^2\ 2s^2\ 2p^6\ 3s^2\ 3p^6\ 3d^5\ 4s^1$

> You could show 4s before 3d if you prefer: $1s^2\ 2s^2\ 2p^6\ 3s^2\ 3p^6\ 4s^2\ 3d^1$.
> Chromium and copper are Period 3 d-block elements with $4s^1$ rather than $4s^2$.

   (ii) a Cr^{3+} ion. **(1 mark)**

$1s^2\ 2s^2\ 2p^6\ 3s^2\ 3p^6\ 3d^3$

> Remember that electrons are first removed from the 4s orbital when d-block ions form.

   (iii) Explain why a chromium(III) compound is coloured. **(3 marks)**

Its d-orbitals are split in energy by ligands. This allows it to absorb visible light so that electron transitions happen from a lower energy to a higher energy.

> Zinc and scandium are d-block elements but they do not form coloured compounds. The d-orbitals of Zn^{2+} ions are all full, so d-d transitions cannot happen, and Sc^{3+} ions have no occupied d-orbitals.

   (iv) Explain why zinc is classified as a d-block element but not as a transition element. **(2 marks)**

Zinc is a d-block element because it has electrons in 3d-orbitals. It is not a transition metal because it does not form a stable ion with an incompletely filled d-orbital.

> Remember that scandium is also a d-block element that is not a transition metal. Using abbreviated electronic configurations:
> • Zn^{2+} is $[Ar]\ 3d^{10}$
> • Sc^{3+} is $[Ar]$

(b) $[Co(H_2O)_6]^{2+}$ ions can react with $NH_2CH_2CH_2NH_2$ to form $[Co(NH_2CH_2CH_2NH_2)_3]^{2+}$ ions.

   (i) Write an equation to represent the change described and state the type of reaction involved.

$[Co(H_2O)_6]^{2+} + 3NH_2CH_2CH_2NH_2$
$\rightarrow [Co(NH_2CH_2CH_2NH_2)_3]^{2+} + 6H_2O$

Ligand exchange.

> Notice that in this reaction a small, uncharged monodentate ligand (H_2O) is substituted by a bidentate ligand.
> There is a large positive increase in the entropy of the system, ΔS_{system}, leading to the formation of a more stable complex.
> An even greater increase occurs when there is substitution by a multidentate ligand such as $EDTA^{4-}$.

   (ii) Give the IUPAC name for $NH_2CH_2CH_2NH_2$. **(1 mark)**

1,2-diaminoethane.

> 1,2-diaminoethane is commonly called ethylenediamine, which is why it is often abbreviated to en.

   (iii) Explain the meaning of the term bidentate ligand. **(2 marks)**

Ligands are species that can donate a pair of electrons to a metal ion. A bidentate ligand can donate two pairs of electrons, forming two dative bonds.

> Remember that the coordination number is the number of dative bonds formed and this is not necessarily the number of ligands bonded to the metal ion.
> There are three en ligands in the ion in part (i) but the coordination number is 6.

(c) State the feature of transition metals that allow them to act as catalysts. **(1 mark)**

They have a variable oxidation number.

> Remember that homogeneous catalysts are in the same phase as the reactants, while heterogeneous catalysts are in a different phase from the reactants.

Measuring reaction rates

Rate of reaction

The **rate of reaction** is the change in concentration of a product or a reactant per unit of time:

$$\text{rate} = \frac{\text{change in concentration of product}}{\text{time}}$$

$$\text{rate} = -\frac{\text{change in concentration of reactant}}{\text{time}}$$

Rate of reaction is always positive (note the minus sign in the second equation). It is measured in different units, depending upon the method used.

Experimental methods

You can choose from a range of methods to measure a rate of reaction. For example:

 volume of gas evolved

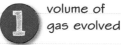 change in mass

 titration

④ colorimetry

You might need a different method from these for a particular reaction. For example:
- the change in pH
- the change in electrical conductivity.

Mass or volume?

To measure the rate of this reaction:

$CaCO_3(s) + 2HCl(aq) \rightarrow CaCl_2(aq) + H_2O(l) + CO_2(g)$

- you could measure the volume of CO_2 made.

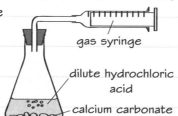

gas syringe

dilute hydrochloric acid

calcium carbonate

Since CO_2 is a dense gas, you could also:

- measure the mass of the reaction mixture as the gas escapes.

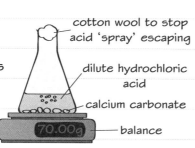

cotton wool to stop acid 'spray' escaping

dilute hydrochloric acid

calcium carbonate

70.00g

balance

The disappearing cross

The reaction between sodium thiosulfate solution and dilute hydrochloric acid is often used in rates of reaction investigations:

$Na_2S_2O_3(aq) + 2HCl(aq)$
$\rightarrow 2NaCl(aq) + H_2O(l) + SO_2(g) + S(s)$

You <u>cannot</u> measure the volume of SO_2 produced because it is very soluble and little escapes.

You <u>can</u> follow the appearance of the sulfur precipitate using the 'disappearing cross' method:

look at cross through the solution

add dilute acid and start timing

sodium thiosulfate solution

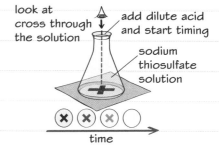

ⓧ ⓧ ⓧ ◯

time

Worked example

Which method would be most suitable when investigating the rate of this reaction? **(1 mark)**

$NaOH(aq) + CH_3COOCH_2CH_3(aq)$
$\rightarrow CH_3CH_2OH(aq) + CH_3COONa(aq)$

☐ **A** Colorimetry
☐ **B** Measuring the change in volume
☐ **C** Measuring the change in mass
☒ **D** Quenching followed by titrating with acid

The reaction is quenched by adding water. The amount of NaOH remaining is found by titration.

Colorimetry

A **colorimeter** is a device that measures the light absorbance of a solution. It is particularly useful where a solution is coloured, such as $MnO_4^-(aq)$.

Unlike the disappearing cross method, which gives an average value for the rate of a reaction, **colorimetry** is a continuous monitoring method.

Now try this

State a suitable method, involving a product, to measure the rate of this reaction:

$Mg(s) + 2HCl(aq) \rightarrow MgCl_2(aq) + H_2(g)$. Explain your choice of method.   **(3 marks)**

Rate equation and initial rate

A rate equation shows the relationship between the rate of a reaction and the reactant concentrations.

Rate equation

In general, rate = $k[A]^m[B]^n$, where:

- k is the **rate constant**
- [A] and [B] are the concentrations of reactants A and B in $mol\,dm^{-3}$
- m and n are 0, 1 or 2.

The **order** with respect to a reactant is the power to which that reactant's concentration is raised.

In the example above, m is the order with respect to reactant A and n is the order with respect to B.

The **overall order of reaction** is the sum of all the individual orders.

Units for the rate constant

The units for k depend upon the units for rate of reaction and the overall order of reaction.

For example, if rate is measured in $mol\,dm^{-3}\,s^{-1}$:

Overall order	Units for k
0	$mol\,dm^{-3}\,s^{-1}$
1	s^{-1}
2	$mol^{-1}\,dm^3\,s^{-1}$

The units are $(mol\,dm^{-3})^{(1-overall\;order)}$ with s^{-1}.
For example, 4th order units are $mol^{-3}\,dm^9\,s^{-1}$.

Initial rate method

You cannot work out the rate equation from the chemical equation for a reaction — instead, you must carry out one or more experiments.

In the initial rate method, you:

- carry out separate experiments for a reaction
- use a different initial concentration of one reactant each time
- compare the rates of the reactions to work out the orders of reaction.

The rate constant and temperature

The rate constant, k, is constant for a given reaction at a given temperature. It changes if:

- the temperature changes
- you add a catalyst or change the catalyst.

This is why you need a constant temperature when you do rate experiments to determine k.

You could also plot a **rate–concentration** graph to work out the order of reaction with respect to a given reactant (see page 117).

Worked example

The table shows the results of four experiments carried out at the same temperature to investigate the rate of reaction between three substances P, Q and R. The reaction is zero order with respect to R.

Experiment	[P]/mol dm^{-3}	[Q]/mol dm^{-3}	[R]/mol dm^{-3}	Initial rate/mol dm^{-3} s^{-1}
1	0.25	0.30	0.10	1.50×10^{-3}
2	0.50	0.30	0.10	3.00×10^{-3}
3	0.25	0.60	0.10	6.00×10^{-3}
4	0.25	0.30	0.20	1.50×10^{-3}

(a) State, with reasons, the orders with respect to reactants P and Q. **(4 marks)**

First order with respect to P because if you compare experiments 1 and 2, [P] doubles while [Q] and [R] stay the same and the rate doubles.

Second order with respect to Q because if you compare experiments 1 and 3, [Q] doubles while [P] and [R] stay the same and the rate goes up × 4.

(b) Write the rate equation. **(1 mark)**

rate = $k[P][Q]^2$

rate = $k[P][Q]^2[R]^0$ is also acceptable ($[R^0] = 1$)

(c) Use the data for experiment 1 to calculate the value for the rate constant. **(3 marks)**

$$k = \frac{rate}{[P][Q]^2} = \frac{1.50 \times 10^{-3}}{(0.25 \times 0.30^2)} = 0.067\,mol^{-2}\,dm^6\,s^{-1}$$

Maths skills You need to be able to solve algebraic equations such as the one above.

Now try this

(a) Use the data above to show that the reaction is zero order with respect to R. **(1 mark)**

(b) Calculate the initial rate if Experiment 3 was repeated but an equal volume of water was added to the reaction mixture. **(3 marks)**

Rate equation and half-life

An order of reaction with respect to a reactant can be determined using a continuous method.

 Practical skills   **Continuous method**

In a continuous rate method to find the order of reaction with respect to a reactant, you first:

 carry out a single experiment for a reaction

2 determine the concentration of a reactant in a sample at different times

3 then plot a **concentration–time graph**.

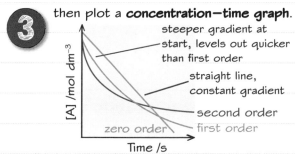

steeper gradient at start, levels out quicker than first order

straight line, constant gradient

second order
first order
zero order

You then:

 calculate the rate at different concentrations

5 plot a **rate–concentration graph**.

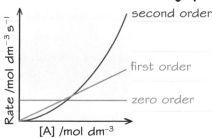

second order
first order
zero order

For zero and first orders, the gradient gives the order of reaction with respect to the reactant. For second order, a graph of rate against $[A]^2$ gives a straight line passing through the origin.

Maths skills   Instantaneous rate of reaction

You can find the **instantaneous** rate of reaction at a given time by:
- plotting a **concentration–time graph** (or **volume–time graph** if you have measured the volume of gas made in the reaction)
- drawing the tangent to the curve at that time
- calculating the tangent's gradient.

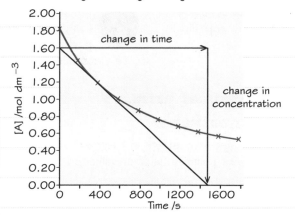

change in time

change in concentration

In the example above at 400 s:

$$\text{rate} = -\frac{(0.00 - 1.60)\,\text{mol dm}^{-3}}{(1470 - 0)\,\text{s}}$$
$$= -\frac{(-1.60)}{1470}$$
$$= 1.09 \times 10^{-3}\,\text{mol dm}^{-3}\,\text{s}^{-1}$$

Half-life of first order reactions

The **half-life**, $t_{\frac{1}{2}}$, of a reaction is:
- the time taken
- for the concentration of the reactant to fall
- to one half of its initial value.

For first order reactions, half-life is:

 constant

2 independent of the initial concentration of the reactant.

In second order reactions, $t_{\frac{1}{2}}$ increases with time and depends on the initial concentration.

Worked example

The graph shows how the concentration of a reactant, A, changes with time during a reaction.

Determine the half-life for the reaction. **(1 mark)**

The half-life is 100 s.

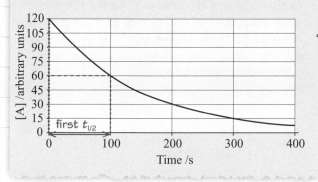

first $t_{1/2}$

The second half-life (60 to 30 units) and the third half-life (30 to 15 units) are both also 100 s.

Now try this

Use the expression $k = 0.693/t_{\frac{1}{2}}$ to calculate the value for k in the Worked Example. **(2 marks)**

Rate-determining steps

Orders of reaction can give information about the rate-determining step in a reaction mechanism.

Hydrolysis of tertiary halogenoalkanes

Halogenoalkanes can be hydrolysed in alkaline solution to produce alcohols (see page 58 for details).

There are two steps in the mechanism for the hydrolysis of tertiary halogenoalkanes.

For example, 2-bromo-2-methylpropane reacts with OH^- ions to form 2-methylpropan-2-ol.

Step I is much slower than Step 2, so it is the **rate-determining step**. Even if the rate of Step 2 is increased, there is little change to the overall rate.

This type of mechanism is an S_N1 mechanism:

 Substitution nucleophilic

 I = unimolecular (only one reactant particle is in the rate-determining step).

$$\text{Step I} \quad H_3C-\underset{\underset{CH_3}{|}}{\overset{\overset{CH_3}{|}}{C}}-Br \underset{slow}{\rightleftharpoons} H_3C-\underset{\underset{CH_3}{|}}{\overset{\overset{CH_3}{|}}{C}}+ + :Br^-$$

$$\text{Step 2} \quad H_3C-\underset{\underset{CH_3}{|}}{\overset{\overset{CH_3}{|}}{C}}+ + :\overset{-}{O}H \xrightarrow{fast} H_3C-\underset{\underset{CH_3}{|}}{\overset{\overset{CH_3}{|}}{C}}-OH$$

Hydrolysis of primary halogenoalkanes

There are also two steps in the mechanism for the hydrolysis of primary halogenoalkanes.

For example, bromoethane reacts with OH^- ions to form ethanol. Again, Step I is much slower than Step 2, so it is the **rate-determining step**.

$$HO^-: + \underset{H}{\overset{H_3C}{\underset{|}{C}}}-Br \xrightarrow{slow} \left[HO---\underset{H\ H}{\overset{H_3C}{\underset{|}{C}}}---Br \right]^- \xrightarrow{fast} HO-\underset{H}{\overset{CH_3}{C}} + :Br^-$$

This type of mechanism is an S_N2 mechanism:

 Substitution nucleophilic

 2 = bimolecular (two reactant particles are in the rate-determining step).

Using order of reaction

In the S_N1 reaction mechanism above
$(CH_3)_3CBr + OH^- \rightarrow (CH_3)_3COH + Br^-$
$rate = k[(CH_3)_3CBr]$

The reaction is first order with respect to $(CH_3)_3CBr$. It is zero order with respect to OH^- (they are in the chemical equation but not in the rate equation). You would expect this because:

✓ Step I is the rate-determining step, so the overall rate depends on the concentration of the reactant involved, $(CH_3)_3CBr$.

✓ Step 2 is much faster, so the overall rate does not depend on OH^- (the other reactant).

In the S_N2 reaction mechanism on the left
$rate = k[CH_3CH_2Br][OH^-]$

It is first order with respect to both reactants and both are involved in the rate-determining step.

Worked example

Iodine and propanone react in the presence of an acid catalyst: $CH_3COCH_3 + I_2 \rightarrow CH_3COCH_2I + HI$

The rate equation is: $rate = k[CH_3COCH_3][H^+]$

(a) How do you know that H^+ ions act as a catalyst in the reaction? **(1 mark)**

H^+ ions appear in the rate equation but not in the overall equation for the reaction.

The rate equation might also be presented as:
$rate = k[CH_3COCH_3]^1[H^+]^1[I_2]^0$

(b) Explain how you know that there are at least two steps in the reaction mechanism. **(2 marks)**

Propanone and H^+ ions are involved in a slow step, as the reaction is first order with respect to both. Iodine must be involved in a later, faster step, as the reaction is zero order with respect to it.

Now try this

Nitrogen dioxide reacts with carbon monoxide: $NO_2 + CO \rightarrow NO + CO_2$ $rate = k[NO_2]^2$
Explain how many molecules of NO_2 and CO are likely to be involved in the rate-determining step. **(2 marks)**

Finding the activation energy

The activation energy for a reaction can be found using experimental data and the Arrhenius equation.

Arrhenius equation

The **Arrhenius equation** links the rate constant. k, to the activation energy, E_a, for a reaction:

$$k = Ae^{-E_a/RT}$$

In the equation:

- A is a measure of the rate at which particles collide (the 'frequency factor').
- R is the **gas constant**, $8.31\,J\,K^{-1}\,mol^{-1}$.
- T is the absolute temperature in kelvin, K.

You will be given this equation (or the one opposite) in the examination if you need it.

Note that:

 1 E_a/RT has no units (as these all cancel out)

 2 k and A have the same units.

Maths skills — Rearranging the Arrhenius equation

Start with the Arrhenius equation:

$$k = Ae^{-E_a/RT}$$

Take natural logarithms, ln, of both sides:

$$\ln k = \frac{-E_a}{R} \times \left(\frac{1}{T}\right) + \ln A$$

This equation follows the general equation for a straight line graph:

$$y = mx + c$$

gradient — intercept on vertical axis, a constant

A graph of $\ln k$ against $1/T$ gives:

 1 a straight line with gradient $-E_a/R$

 2 an intercept on the $\ln k$ axis equal to $\ln A$.

Worked example

Peroxydisulfate ions and iodide ions react together in solution: $S_2O_8^{2-}(aq) + 2I^-(aq) \rightarrow 2SO_4^{2-}(aq) + I_2(aq)$

The table shows some data about the reaction, determined using experimental data.

Temperature, T/K	Rate constant, k	$\ln k$	$1/T$/K^{-1}
300	5.13×10^{-3}	-5.27	3.33×10^{-3}
310	8.33×10^{-3}	-4.79	3.23×10^{-3}
320	1.28×10^{-2}	-4.36	3.13×10^{-3}
330	2.01×10^{-2}	-3.91	3.03×10^{-3}
340	3.01×10^{-2}	-3.50	2.94×10^{-3}

(a) Use the data to plot a graph of $\ln k$ against $1/T$. Draw a line of best fit through the points. **(2 marks)**

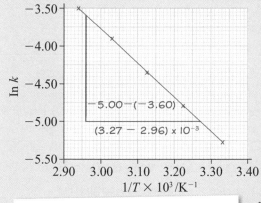

Be careful! Plot the points carefully because the examiners will check their positions. They will also check that your plotted points occupy more than half of the grid. If you expect the line to be straight, as here, use a 30 cm ruler to draw the line of best fit. Do not join the points with separate lines.

(b) Calculate the gradient of the line. **(2 marks)**

change in $\ln k = -5.00 - (-3.60) = -1.40$
change in $1/T = (3.27 - 2.96) \times 10^{-3} = 0.31 \times 10^{-3}$
gradient $= -1.40 \div (0.31 \times 10^{-3}) = -4516$

(c) Use your answer to part (b) to calculate the activation energy, E_a, for the reaction. **(2 marks)**

$-E_a/R = -4516$
$E_a = 4516 \times 8.31 = +37500\,J\,mol^{-1} = +37.5\,J\,mol^{-1}$

Activation energies are always positive, so make sure you include the + sign in your answer.

Now try this

The rate constant for a reaction was determined at different temperatures.
(a) Calculate the values of (i) $1/T$ where $T = 298\,K$ (ii) $\ln k$ where $k = 3.60 \times 10^{-3}$ **(2 marks)**
(b) The gradient of the graph of $\ln k$ against $1/T$ was -3250. Calculate E_a in $kJ\,mol^{-1}$. **(2 marks)**

Exam skills 11

This exam-style question uses knowledge and skills you have already revised. Have a look at pages 115–117 for a reminder about **reaction rates** and **order of reaction**.

Worked example

Hydrogen peroxide reacts with iodide ions to form iodine. If thiosulfate ions, $S_2O_3^{2-}$, are also present, they react with the iodine formed. Once all these ions have reacted, iodine is no longer reduced. The appearance of iodine is detected by starch solution.

(a) Describe the final colour of the reaction mixture.

Blue-black.

(b) The concentration of I⁻ ions was varied while keeping the concentrations and volumes of the other reagents the same and the time for the mixture to change colour was recorded.

[I⁻]/ mol dm⁻³	time/ s	1/time/ s⁻¹
0.040	16.5	0.0606
0.030	22.4	0.0446
0.016	41.7	0.0239
0.008	85.2	0.0117

(i) Complete the table, and plot a graph of 1/time on the vertical axis against [I⁻]. **(3 marks)**

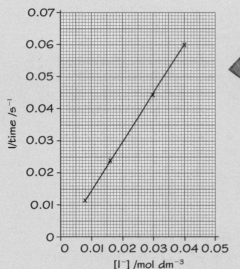

(ii) 1/time is a measure of the initial rate of reaction. Deduce the order of reaction with respect to iodide ions, and justify your answer. **(2 marks)**

The reaction is first order with respect to iodide ions because the rate is proportional to [I⁻]. The graph is a straight line (it has a constant gradient).

(iii) Describe a way in which the experiment could be improved, without changing the method, measuring apparatus or the solutions used. **(1 mark)**

You could repeat the experiment.

This describes a version of the 'iodine clock experiment':

1. $H_2O_2 + 2H^+ + 2I^- \rightarrow I_2 + 2H_2O$
2. $I_2 + 2S_2O_3^{2-} \rightarrow S_4O_6^{2-} + 2I^-$

Sulfuric acid is added to provide H⁺ ions.

Practical skills Water is usually added before the reactants are mixed together so that the total volume remains the same.

Laboratory digital stop clocks often time to ±0.01 s, but this precision is unnecessary when you have to judge when to stop the clock.

Practical skills Make sure you can recall colours and colour changes from practical activities. 'Purple' would not be correct here.

When you plot a graph, you need to make sure you:
- choose sensible scales on which the plotted points occupy at least half of the grid supplied
- use linear scales
- plot points accurately
- draw a line of best fit.

You should also:
- make sure you have your axes the right way round
- label each axis with the quantity and unit.

Command word: Deduce

If a question asks you to **deduce** something, it means you need to reach a conclusion from the information given.

Command word: Justify

If a question asks you to **justify** something, it means you need to give evidence to prove something.

Practical skills Since temperature is one of the factors that determines the rate of a reaction, the experiment should be repeated at the same temperature. This could be done using a thermostatic water bath.

Identifying aldehydes and ketones

Aldehydes and ketones contain the carbonyl group, C=O.

The carbonyl group in aldehydes

In **aldehydes**, the carbonyl group is located at carbon atom 1. For example, this is pentanal:

Their names end in -al and you show the functional group as CHO in their structural formulae.

Name	Structural formula
Methanal	HCHO
Ethanal	CH_3CHO
Propanal	CH_3CH_2CHO
2-methylpropanal	$(CH_3)_2CHCHO$

The carbonyl group in ketones

In ketones, the carbonyl group is not at the end of the molecule. For example, this is pentan-2-one:

Their names end in -one and you show the functional group as CO in their structural formulae.

Name	Structural formula
Propanone	CH_3COCH_3
Butanone	$CH_3COCH_2CH_3$
Pentan-2-one	$CH_3COCH_2CH_2CH_3$
Pentan-3-one	$CH_3CH_2COCH_2CH_3$

Hydrogen bonds

The **intermolecular forces** between the molecules of aldehydes and ketones include:

- London forces
- permanent dipole–dipole forces (due to their polar C=O bonds).

The boiling temperatures of aldehydes and ketones increase as their molar masses increase and the London forces increase (see page 14).

Methanal is a gas at room temperature but other lower carbonyl compounds are liquids.

Carbonyl compounds:

- do not form hydrogen bonds between their molecules
- do form hydrogen bonds with water, so lower aldehydes and ketones are soluble in water.

Oxidation reactions

Aldehydes can be oxidised using oxidising agents.

In general: $XCHO + [O] \rightarrow XCOOH$ —— carboxylic acid
oxidising agent

Ketones are difficult to oxidise. They give negative results with these tests for aldehydes:

Reagent	Observed change
Dichromate(VI) ions acidified with dilute H_2SO_4	orange solution → green solution
Fehling's solution	blue solution → red precipitate
Benedict's solution	blue solution → red precipitate
Tollens' reagent	colourless solution → silver mirror

Worked example

Two isomers of C_3H_6O, isomer A and isomer B, both contain a carbonyl group.

(a) Describe the expected observations when 2,4-dinitrophenylhydrazine is added to them. **(1 mark)**

An orange–yellow precipitate forms.

(b) Describe how you could identify each carbonyl compound using the products in part (a). **(3 marks)**

Recrystallise the products to purify them, and then dry them. Measure their melting temperatures and compare them with values from a data book.

2,4-dinitrophenylhydrazine (2,4-DNPH or Brady's reagent) is used as a qualitative test for the presence of a carbonyl group. It tends to form orange–yellow precipitates with aliphatic carbonyl compounds and orange–red precipitates with aromatic carbonyl compounds.

Different 2,4-DNPH derivatives (the precipitates formed) have different melting temperatures. You can identify the original carbonyl compound by matching the melting temperature of its derivative.

Now try this

Other than using Brady's reagent, describe a simple laboratory test to distinguish between the two carbonyl compounds described in the Worked Example. **(3 marks)**

Optical isomerism

Optical isomerism results from a chiral centre in a molecule with an asymmetric carbon atom.

Chiral centres

A **chiral centre** is an atom in a molecule that results in optical isomers. In organic molecules, you are looking for an **asymmetric** carbon atom — one that is bonded to four different atoms or groups. For example:

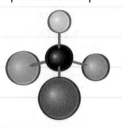 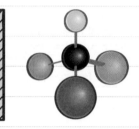

These two structures are:

1 an object and its mirror image

2 non-superimposable (no matter how you turn them around, they will not be identical)

3 stereoisomers called **enantiomers**.

Plane-polarised light

Visible light is electromagnetic radiation. Its waves:

✓ are transverse (their vibrations are at 90° to the direction that the wave travels).

In ordinary light, such as from a lamp, the vibrations are in all directions or **planes**.

Polarising materials, such as Polaroid, only let light with vibrations in one plane through. The light that leaves the other side is **plane-polarised** light.

In a **polarimeter**, when two Polaroid sheets are at 90° to each other, no light leaves the second one.

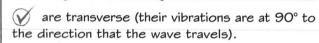
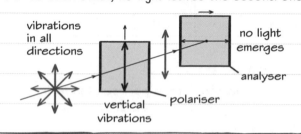

vibrations in all directions

no light emerges

analyser

vertical vibrations

polariser

Enantiomers

The two enantiomers of a given compound have similar properties, except:

- they **rotate** the **plane of polarisation** of plane-polarised light in opposite directions.

Each one rotates the plane by the same amount, so one might rotate it by +10° and the other by −10°.

A **racemic mixture** contains equal amounts of the two enantiomers — it does not rotate the plane of plane-polarised light.

Polarimetry

Polarimetry measures the **angle of rotation**:

- The analyser is rotated to maximum darkness.
- A solution of an optically active substance is put in a tube between the polariser and analyser.
- The analyser is rotated again until maximum darkness is obtained.
- The angle of rotation is measured.

Monochromatic (single frequency) light gives a more accurate reading than white light.

Worked example

Lactic acid, 2-hydroxypropanoic acid, is a compound found in sour milk and in muscles.
Explain why lactic acid is chiral. Indicate the feature involved on the structure below. **(3 marks)**

It has an asymmetric carbon atom (circled) which has four different groups attached, so the molecule is non-superimposable on its mirror image.

Lactic acid in muscles. It is the **dextrorotatory** or (+) enantiomer, which rotates plane-polarised light clockwise.

Sour milk contains a racemic mixture, an equal amount of the **laevorotatory** or (−) enantiomer.

The molecule could also have been described as a three-dimensional molecule with no plane of symmetry.

Now try this

Ethanal reacts with hydrogen cyanide (via an alkaline solution of potassium cyanide) to produce 2-hydroxypropanenitrile (shown):

H_3C—C—CN

Page 12 shows how you draw such diagrams.

Draw three-dimensional diagrams to show the shapes of the two enantiomers of 2-hydroxypropanenitrile. **(2 marks)**

Optical isomerism and reaction mechanisms

When an optically active organic compound undergoes a nucleophilic substitution reaction, the formation of an optically inactive product is evidence of an S_N1 mechanism, while the formation of an optically active product is evidence of an S_N2 mechanism.

Evidence for S_N1 mechanisms

If an optically active organic reactant produces a racemic mixture of products, this is evidence for an S_N1 reaction mechanism.

For example, 1-bromo-1-fluoroethane has a chiral carbon atom and is optically active.

It can react with hydroxide ions in a nucleophilic substitution reaction to form 1-fluoroethanol.

This product is not optically active.

The OH^- ion can approach from above or below the trigonal planar carbocation intermediate.

Evidence for S_N2 mechanisms

If an optically active organic reactant produces an optically active product, this is evidence for an S_N2 reaction mechanism.

For example, look again at 1-bromo-1-fluoroethane but this time showing an S_N2 reaction mechanism.

Worked example

An optically active isomer of 2-bromobutane reacts with hydroxide ions to form butan-2-ol. Explain why this product is a racemic mixture. **(2 marks)**

The reaction involves a planar carbocation intermediate in an S_N1 reaction mechanism. The hydroxide ion can attack from above and below.

Maths skills Representing three-dimensional objects using two-dimensional diagrams is an important maths skill.

Page 118 explains how you can use rate equations to provide evidence for rate-determining steps and for S_N1 and S_N2 reaction mechanisms.

Now try this

An optically active isomer of 2-chloro-2-methylbutane reacts with hydroxide ions to form 2-methylbutan-2-ol. Why is this product optically active? **(1 mark)**

☐ **A** The reaction is nucleophilic substitution.

☐ **B** 2-chloro-2-methylbutane forms a carbocation intermediate.

☐ **C** 2-chloro-2-methylbutane forms a five-bonded intermediate.

☐ **D** 2-chloro-2-methylbutane contains a chiral carbon atom.

Reactions of aldehydes and ketones

Carbonyl compounds undergo nucleophilic addition reactions with CN⁻ ions to form hydroxynitriles.

Reactions with iodine

Triiodomethane, CHI_3, is an insoluble yellow solid. It forms in reactions between certain carbonyl compounds and an alkaline solution of iodine.

CHI_3 ('iodoform') forms when this group is present:

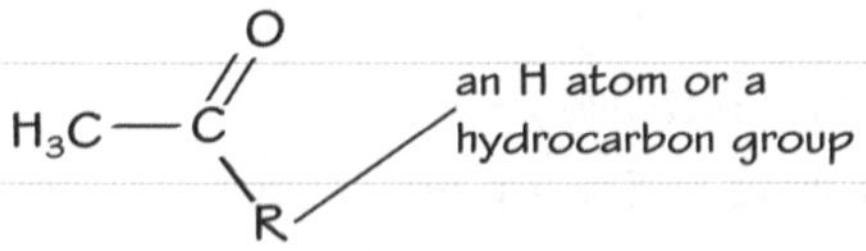

an H atom or a hydrocarbon group

In general:

$CH_3COR + 3I_2 + 4OH^-$
$\rightarrow CHI_3 + RCOO^- + 3I^- + 3H_2O$

Some alcohols give positive results:

Type of alcohol	Formation of CHI_3
Primary	✗ unless ethanol
Secondary	✓ if R = $CH(CH_3)OH$
Tertiary	✗

This means that:
- ethanal is the only aldehyde to produce CHI_3
- ketones only produce CHI_3 if they have a methyl group attached to the CO group.

For example, pentan-2-one gives a positive result but pentan-3-one does not.

Reaction with hydrogen cyanide

Carbonyl compounds react with hydrogen cyanide, HCN, in the presence of potassium cyanide, KCN.

The KCN increases the concentration of cyanide ions, CN⁻, allowing the reaction to proceed more quickly.

The organic products are hydroxynitriles, compounds containing an −OH group and a −CN group.

The table shows two carbonyl compounds and the hydroxynitriles they form.

An aldehyde

propanal + HCN → 2-hydroxybutanenitrile

A ketone

propanone + HCN → 2-hydroxy-2-methylpropanenitrile

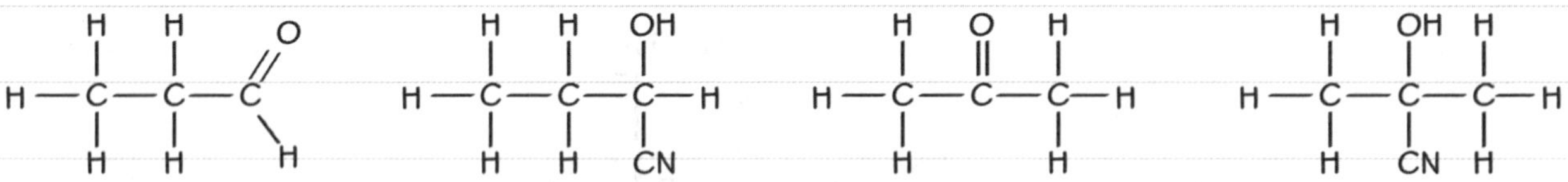

The general reaction mechanism is shown below. It is an example of a **nucleophilic addition** reaction.

The product contains one more C atom than the reactant.

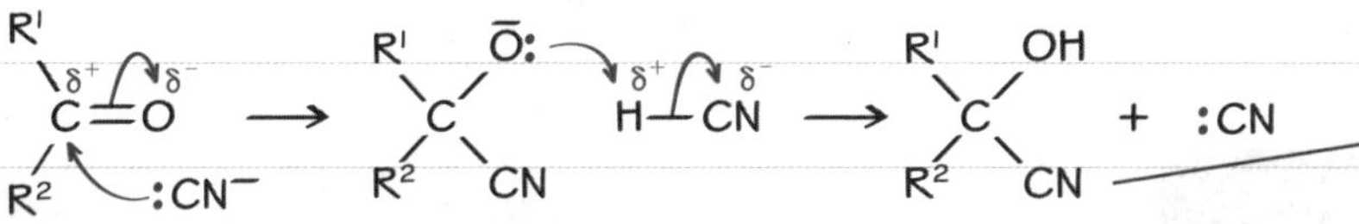

Notice that 2-hydroxybutanenitrile has a **chiral centre** (2-hydroxy-2-methylpropanenitrile does not).

It is produced as a racemic mixture. Equal amounts of its two **enantiomers** are made because:
- The arrangement of the two groups attached to the carbonyl group is **planar**.
- The **nucleophile** (the CN⁻ ion) has an equal chance of attacking from above or below the plane.

Worked example

Lithium tetrahydridoaluminate, $LiAlH_4$, is a reducing agent. It can reduce carbonyl compounds to alcohols when dissolved in dry ether.

Write an equation to show the reduction of propanal and name the product. **(2 marks)**

$CH_3CH_2CHO + 2[H] \rightarrow CH_3CH_2CH_2OH$
The product is propan-1-ol.

$LiAlH_4$ is also called lithium aluminium hydride. You represent the reducing agent as [H] in the equations — there is no need to write $LiAlH_4$:
- Aldehydes are reduced to primary alcohols.
- Ketones are reduced to secondary alcohols.

Now try this

An unbranched carbonyl compound contains four carbon atoms. It produces a yellow precipitate in an alkaline solution of iodine. Name this carbonyl compound. Predict the organic products of its reaction with $LiAlH_4$ in dry ether, and with HCN in the presence of KCN. **(3 marks)**

Carboxylic acids

Carboxylic acids turn blue litmus paper red, and react with carbonates to release carbon dioxide gas.

The carboxyl group

The **carboxyl** group consists of:
- a carbonyl group, C=O
- a hydroxyl group, −OH

$$R-C{\overset{\displaystyle O}{\underset{\displaystyle O-H}{\|}}}$$

These are on the same C atom. They influence each other and carboxylic acids have some properties not shared with carbonyl compounds or alcohols.

The carbonyl group is located at carbon atom 1. For example, this is 3-methylbutanoic acid:

$$H-\overset{H}{\underset{H}{C}}-\overset{H}{\underset{CH_3}{C}}-\overset{H}{\underset{H}{C}}-C{\overset{O}{\underset{OH}{/\!/}}}$$

Their names end in-anoic acid and you show the functional group as COOH in structural formulae.

Name	Structural formula
Methanoic acid	HCOOH
Ethanoic acid	CH_3COOH
Propanoic acid	CH_3CH_2COOH

Physical properties

The **intermolecular forces** between carboxylic acid include:

☑ London forces

☑ permanent dipole−dipole forces (due to their polar C=O, C−O and O−H bonds)

☑ hydrogen bonds (unlike carbonyl compounds).

$$R-\overset{H}{\underset{H}{C}}-\overset{\delta+}{C}{\overset{:\overset{\delta-}{O}:----H-\overset{\delta-}{O}}{\underset{\overset{\delta-}{O}-H----:\overset{\delta-}{O}}{}}}\overset{\delta+}{C}-\overset{H}{\underset{H}{C}}-R$$

hydrogen bonds cause **dimers** to form

Carboxylic acids with fewer than about ten carbon atoms are liquids at room temperature:
- their boiling temperatures increase as their molar masses increase, and the London forces increase (see page 14).

Carboxylic acids form hydrogen bonds with water. Solubility decreases with increasing chain length.

Making carboxylic acids

Carboxylic acids can be made in the lab by:

 Oxidation of primary alcohols or aldehydes using $K_2Cr_2O_7$ acidified with dilute sulfuric acid, e.g. for propanoic acid:

$$CH_3CH_2CH_2OH + 2[O] \rightarrow CH_3CH_2COOH + H_2O$$
propan-1-ol

$$CH_3CH_2CHO + [O] \rightarrow CH_3CH_2COOH$$
propanal

 Hydrolysis of nitriles under reflux with dilute hydrochloric acid, e.g. from propanenitrile:

$$CH_3CH_2CN + H^+ + 2H_2O \rightarrow CH_3CH_2COOH + NH_4^+$$

Reactions of carboxylic acids

The reactions of carboxylic acids include:

1 Reduction to primary alcohols using $LiAlH_4$ in dry ether, e.g. for propanoic acid:

$$CH_3CH_2COOH + 4[H] \rightarrow CH_3CH_2CH_2OH + H_2O$$
propan-1-ol

2 Chlorination to produce **acyl chlorides** by reaction with anhydrous PCl_5, e.g.

$$CH_3CH_2COOH + PCl_5$$
$$\rightarrow CH_3CH_2COCl + POCl_3 + HCl$$
propanoyl chloride   removed by distillation   produces misty fumes

Worked example

Describe what you would expect to see in the reaction between dilute ethanoic acid and sodium carbonate solution. Include an equation. **(3 marks)**

The two substances are colourless. There is brief effervescence when they are mixed, due to the production of carbon dioxide.

$$2CH_3COOH + Na_2CO_3 \rightarrow 2CH_3COONa + H_2O + CO_2$$

The organic product, sodium ethanoate, can also be made by reaction with sodium hydroxide solution:
$$CH_3COOH + NaOH \rightarrow CH_3COONa + H_2O$$
The formula for the ethanoate ion is CH_3COO^-.

Now try this

Describe three ways in which 3-methylbutanoic acid may be produced. Name the organic reactant(s) needed, and include equations in your answers. **(6 marks)**

Making esters

Most esters are colourless liquids with distinct smells. They are used as flavourings and solvents.

Making esters

Esters can be made by reacting alcohols with:
- carboxylic acids
- acyl chlorides

For example, ethanoic acid reacts with methanol in the presence of an acid catalyst to make methyl ethanoate, CH_3COOCH_3, and water.

The ester functional group, COO

Acyl chlorides produce hydrogen chloride gas instead of water when they react with alcohols.

Acyl chlorides

The functional group in acyl chlorides has:
- a carbonyl group, C=O
- a chlorine atom, Cl.

The acyl chloride group is located at carbon atom 1. For example, this is 3-methylbutanoyl chloride:

Their names end in -anoyl chloride, and you show the functional group as COCl in structural formulae.

Name	Structural formula
Ethanoyl chloride	CH_3COCl
Propanoyl chloride	CH_3CH_2COCl
Butanoyl chloride	$CH_3CH_2CH_2COCl$

Acyl chlorides react with water to form carboxylic acids and misty fumes of hydrogen chloride. For example, ethanoyl chloride makes ethanoic acid:

$$CH_3COCl + H_2O \rightarrow CH_3COOH + HCl$$

Naming esters

When you name an ester:
- The first part comes from the alcohol.
- The second part comes from the other reactant.

When you write the formula for an ester, you show its functional group as COO.

Name	Structural formula
Ethyl methanoate	$HCOOCH_2CH_3$
Ethyl ethanoate	$CH_3COOCH_2CH_3$
Methyl butanoate	$CH_3CH_2CH_2COOCH_3$

In all these structural formulae, the part contributed by the alcohol is on the right of COO.

Hydrolysis of esters in alkaline conditions

Esters can be **hydrolysed** by heating them under reflux with sodium hydroxide solution. In general:

$$R_1COOR_2 + NaOH \rightarrow R_1COONa + R_2OH$$

For example, methyl ethanoate is hydrolysed to form sodium ethanoate and methanol:

$$CH_3COOCH_3 + NaOH \rightarrow CH_3COONa + CH_3OH$$

You can convert the sodium ethanoate into ethanoic acid by heating it with excess dilute hydrochloric acid (or sulfuric acid):

$$CH_3COONa + HCl \rightarrow CH_3COOH + NaCl$$

Worked example

Describe how you could hydrolyse the ester methyl ethanoate in acidic conditions. Include an equation.

(3 marks)

Add some dilute hydrochloric acid or dilute sulfuric acid, then heat under reflux conditions.

$$CH_3COOCH_3 + H_2O \rightleftharpoons CH_3COOH + CH_3OH$$

Now try this

1 Name the two products of the reaction between pentanoyl chloride and propan-1-ol. Draw the displayed formula for the organic product.

(3 marks)

2 Suggest two reasons why acyl chlorides react more quickly with alcohols than carboxylic acids do. **(2 marks)**

Dilute hydrochloric acid and dilute sulfuric acid can also catalyse the reactions between carboxylic acids and alcohols to produce esters.

Notice that, unlike base-catalysed hydrolysis, the reaction does not go to completion.

However, you do not have to do a further step to produce the carboxylic acid from its salt.

Making polyesters

Polyesters are condensation polymers containing very many ester functional groups.

What are polyesters?

Polyesters are relatively large molecules formed in **condensation reactions** between many smaller **monomer** molecules.

Polyethylene terephthalate or 'PET' is a common polyester used to make:

- drinks bottles
- synthetic fibres
- food packaging.

In general, polyesters are made from:

 a diol (a compound with two hydroxyl groups)

 a dicarboxylic acid (a compound with two carboxyl groups) or a dioyl dichloride (a compound with two acyl chloride groups).

Condensation vs addition

Polymers such as poly(ethene) are **addition polymers** (see page 55 for more details).

In addition polymerisation:

 The monomers are **unsaturated** organic compounds such as alkenes.

 The polymer is the only product formed.

In condensation polymerisation:

 The monomers are organic compounds with at least two reactive groups (such as hydroxyl, carboxyl or amine groups).

 The polymer is not the only product formed (a smaller product such as H_2O or HCl forms too).

Polyethylene terephthalate, PET

The diol monomer for PET is ethane-1,2-diol.

two hydroxyl groups —

$$H-O-\underset{\underset{H}{|}}{\overset{\overset{H}{|}}{C}}-\underset{\underset{H}{|}}{\overset{\overset{H}{|}}{C}}-O-H$$

The other monomer is benzene-1,4-dicarboxylic acid.

two carboxyl groups

benzene ring (see page 128)

A hydroxyl group on the diol can react with a carboxyl group on the dicarboxylic acid.

Notice that the organic product has a hydroxyl group and a carboxyl group.
Further reactions can happen, leading to a long polymer molecule.

Worked example

Polyethylene terephthalate is a polyester made from ethane-1,2-diol and benzene-1,4-dicarboxylic acid. Draw its repeat unit. **(1 mark)**

Benzene-1,4-dicarbonoyl dichloride can be used instead of benzene-1,4-dicarboxylic acid.

The same polymer forms but with HCl, not H_2O.
The reaction is still condensation polymerisation even though water is not the inorganic product.

Now try this

The repeat unit for a polyester is shown on the right.

(a) Name the diol and dicarboxylic acid needed to form this polymer. **(2 marks)**

(b) Name the type of polymerisation, and the inorganic product formed. **(2 marks)**

Benzene

Benzene is a colourless, flammable liquid with the molecular formula C_6H_6.

Kekulé model for benzene

There are over 300 structures with the formula C_6H_6. One of these is the **Kekulé model** for benzene (first proposed in the 19th century).

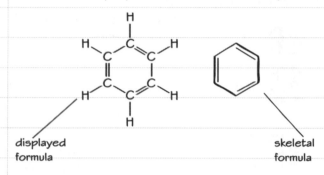

displayed
formula

skeletal
formula

Delocalised model for benzene

In the modern, **delocalised model** for benzene:

• six orbitals overlap
• to form three π bonds.

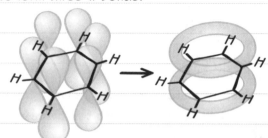

The π bonds form clouds of delocalised electrons above and below the plane of the carbon ring.

Bond length evidence

If the Kekulé model is correct, there will be:

 C=C and C−C bonds.

 Two different carbon−carbon bond lengths.

The measured carbon−carbon bond length in benzene is 0.139 nm. This is intermediate between:

• the C−C bond length (0.154 nm) and
• the C=C bond length (0.134 nm).

This is not predicted by the Kekulé model and is evidence in support of the delocalised model.

It suggests that the carbon−carbon bonds in benzene are identical to one another.

Enthalpy data evidence

Cyclohexene reacts with H_2 to form cyclohexane:

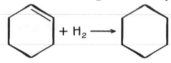

$\Delta H_r^\ominus = -120 \text{ kJ mol}^{-1}$

If the Kekulé model (which contains three C=C bonds, not one) is correct, $\Delta H_r^\ominus$ for the hydrogenation of benzene should be:

$3 \times (-120) = -360 \text{ kJ mol}^{-1}$

The measured value is -208 kJ mol^{-1}. This means:

• benzene is more stable (by 152 kJ mol^{-1}) than the Kekulé model predicts.

The difference is the **delocalisation enthalpy**.

Worked example

(a) Describe what would be seen when benzene burns in air, and write an equation for the complete combustion of benzene. **(2 marks)**

An orange, smoky flame.

$C_6H_6 + 7\frac{1}{2}O_2 \rightarrow 6CO_2 + 3H_2O$

(b) Benzene reacts with bromine under reflux in the presence of a catalyst. Write an equation for the reaction that produces bromobenzene and name a suitable catalyst for the reaction. **(2 marks)**

$C_6H_6 + Br_2 \rightarrow C_6H_5Br + HBr$

Aluminium bromide is a suitable catalyst.

Benzene has a high ratio of carbon atoms to hydrogen atoms. A smoky flame is typical of the combustion of such compounds.

Alkenes readily react with bromine to form colourless dibromo products, but benzene does not.

Iron(III) bromide is an alternative catalyst. Iron filings could be used instead but are not really a catalyst − the iron reacts with bromine to form iron(III) bromide and is permanently changed.

Now try this

1 Explain why the molecule represented by the Kekulé model for benzene may be referred to as cyclohexa-1,3,5-triene. **(2 marks)**

2 How many electrons are involved in the delocalised π bonds in benzene? **(1 mark)**

Halogenation of benzene

Benzene is resistant to bromination as it has delocalised π bonds, whereas alkenes are readily brominated because they have localised electron density in their π bond.

Electrophilic substitution

Benzene undergoes **electrophilic substitution** reactions. In general there are main two steps.

 Two of the electrons in the delocalised π bond form a dative covalent bond with an electrophile. A charged intermediate forms.

 A hydrogen ion, H^+, is eliminated. The H atom attached to the benzene ring was there all the time (remember: C_6H_6).

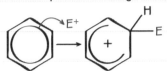

Mechanism for bromination of benzene

Benzene reacts with bromine in the presence of catalyst consisting of $AlBr_3$ or $FeBr_3$:

$$C_6H_6 + Br_2 \rightarrow C_6H_5Br + HBr$$

The **electrophile** is a Br^+ ion generated like this:

$$AlBr_3 + Br_2 \rightarrow [AlBr_4]^- + Br^+$$

Forming the inorganic products

The H^+ ion released in the last step below reacts with the $[AlBr_4]^-$ ion:

$$[AlBr_4]^- + H^+ \rightarrow AlBr_3 + HBr$$

This regenerates the $AlBr_3$ catalyst and produces hydrogen bromide.

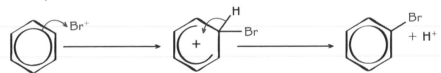

A dative bond forms and the Br^+ ion is added to a carbon atom.

Two electrons from the C—H bond restore the delocalised π bond (the H atom in the diagram was already there).

A hydrogen ion leaves, and bromobenzene forms.

Worked example

Explain why phenol can be brominated at room temperature without a catalyst. **(3 marks)**

Phenol has a hydroxyl group. One of the lone pairs of electrons on the oxygen atom in this group delocalises with the electrons in the delocalised pi bond. This increases the electron density of the ring, activating it and making it more susceptible to electrophilic attack.

Unlike benzene, phenol decolourises orange–brown bromine water (making 2,4,6-tribromophenol).

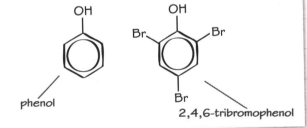

phenol

2,4,6-tribromophenol

Now try this

Benzene reacts with chlorine in the presence of $AlCl_3$ as a catalyst. The reaction happens in a similar way to the reaction between benzene and bromine.

(a) State the type of reaction mechanism involved in these reactions. **(1 mark)**

(b) Write two equations to show how $AlCl_3$ acts as a catalyst in the reaction. **(2 marks)**

(c) Outline the reaction mechanism for the reaction. **(3 marks)**

Nitration of benzene

Nitrobenzene is a toxic pale yellow liquid, insoluble in water.

Nitration

During **nitration**:

- A hydrogen atom is substituted by a **nitro** group, $-NO_2$.
- Benzene, C_6H_6, is converted into **nitrobenzene**, $C_6H_5NO_2$.

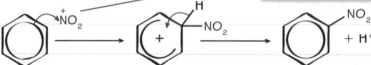

nitronium ion

Nitrating mixture

The nitronium ion, NO_2^+, is the electrophile that attacks the delocalised π bond in benzene.

It is generated using a mixture of:

☑   concentrated nitric acid and

☑   concentrated sulfuric acid.

These react together:

$$HNO_3 + H_2SO_4 \rightarrow H_2NO_3^+ + HSO_4^-$$

then

$$H_2NO_3^+ \rightarrow NO_2^+ + H_2O$$

The overall reaction is:

$$HNO_3 + H_2SO_4 \rightarrow NO_2^+ + HSO_4^- + H_2O$$

Mechanism for nitration of benzene

The mechanism is **electrophilic substitution**.

Be careful! The electrophile is the NO_2^+ ion, not the nitro group, NO_2.

A dative bond forms and the nitronium ion, NO_2^+, is added to a carbon atom.

Two electrons from the C–H bond restore the delocalised π bond (the H atom in the diagram was already there).

A hydrogen ion leaves (and is accepted by HSO_4^-), and nitrobenzene forms.

Note that sulfuric acid acts as a **catalyst** – the H^+ ion and HSO_4^- ion react together to regenerate it:

$$H^+ + HSO_4^- \rightarrow H_2SO_4$$

Worked example

Benzene is nitrated using a mixture of concentrated nitric acid and concentrated sulfuric acid at about 50 °C. Explain the role of these two acids in terms of proton transfer. **(3 marks)**

Protonated nitric acid forms in the reaction:

$$HNO_3 + H_2SO_4 \rightarrow H_2NO_3^+ + HSO_4$$

Sulfuric acid is acting as a Brønsted–Lowry acid because it donates a proton to nitric acid. Nitric acid is acting as a Brønsted–Lowry base because it accepts a proton from sulfuric acid.

Further substitution

Further substitution happens if the temperature is raised above 50°C.

1,3-dinitrobenzene is the major product formed.

Sulfuric acid is a stronger acid than nitric acid.

Now try this

1,3-dinitrobenzene is produced by the nitration of nitrobenzene. The reaction mechanism is similar to the one for the nitration of benzene.

(a) State the type of reaction mechanism involved. **(1 mark)**

(b) Write an equation to show the generation of the nitronium ion. **(1 mark)**

(c) Write the mechanism, including curly arrows, for the nitration of nitrobenzene. **(3 marks)**

Treat the existing nitro group as if it is attached to carbon atom number 1.

Friedel–Crafts reactions

Friedel–Crafts reactions involve aluminium halides, and must be carried out under reflux in anhydrous conditions because the reactants (apart from benzene) react with water.

Alkylation

During **alkylation**, a hydrogen atom is substituted by an **alkyl** group. For example:

- Benzene is converted into methylbenzene if the alkyl group is $-CH_3$.

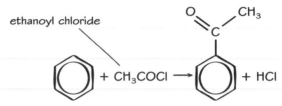

The electrophile for this reaction is the CH_3^+ ion. It is generated from CH_3Cl with an $AlCl_3$ catalyst:

$$CH_3Cl + AlCl_3 \rightarrow [AlCl_4]^- + CH_3^+$$

Acylation

During **acylation**, a hydrogen atom is substituted by an **acyl** group. For example:

- Benzene is converted into phenylethanone if the acyl group is $-COCH_3$.

The electrophile for this reaction is the CH_3CO^+ ion. It is generated from CH_3COCl with an $AlCl_3$ catalyst:

$$CH_3COCl + AlCl_3 \rightarrow [AlCl_4]^- + CH_3CO^+$$

Mechanism for alkylation of benzene

The mechanism is **electrophilic substitution**. The one below is for the formation of methylbenzene.

Be careful! Show the positive charge on the electrophile over the carbon atom where the C—Cl bond was broken.

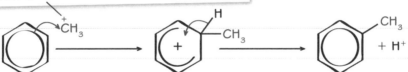

A dative bond forms and the $\overset{+}{C}H_3$ ion is added to a carbon atom.

Two electrons from the C—H bond restore the delocalised π bond (the H atom in the diagram was already there).

A hydrogen ion leaves (which is accepted by $[AlCl_4]^-$), and methylbenzene forms.

The H^+ ion released in the last step above reacts with the $[AlCl_4]^-$ ion and regenerates the catalyst:

$$[AlCl_4]^- + H^+ \rightarrow AlCl_3 + HCl$$

Worked example

Benzene reacts with ethanoyl chloride, in the presence of an aluminium chloride catalyst, to form phenylethanone and hydrogen chloride.
Write the mechanism, including curly arrows, for this reaction. **(3 marks)**

The positive charge on the electrophile is shown where the C—Cl bond was broken.

Now try this

The following compounds can be made from benzene using Friedel–Crafts reactions. For each one, state the name and structural formula of the other organic reactant. Draw the displayed formula of the electrophile it forms.

(a) Propylbenzene   **(3 marks)**   (b) Phenylpropan-1-one   **(3 marks)**

The locant 1 shows that the benzene ring and carbonyl group are attached to the same carbon atom.

Making amines

Primary aliphatic amines such as ethylamine are made from halogenoalkanes or nitriles.

The amine group

You can imagine the amine group as:
- an ammonia molecule, NH_3, in which
- one or more of the H atoms has been replaced
- by an alkyl group (carbon chain) or aryl group (benzene ring).

Primary amine	Secondary amine	Tertiary amine
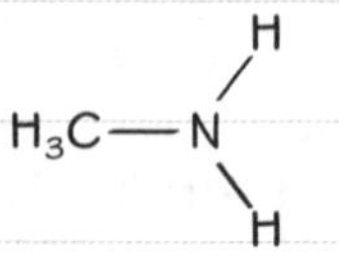 H_3C-N with H (top) and H (bottom)	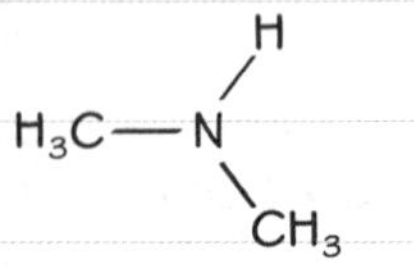 H_3C-N with H (top) and CH_3 (bottom)	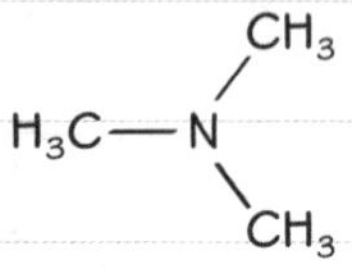 H_3C-N with CH_3 (top) and CH_3 (bottom)
Methylamine	Dimethylamine	Trimethylamine

Primary amines contain the **amino** group, $-NH_2$.

Naming amines

The simplest naming system treats the amino group as having an alkyl or aryl group attached. For example:

Name	Structural formula
Methylamine	CH_3NH_2
Ethylamine	$CH_3CH_2NH_2$
Propylamine	$CH_3CH_2CH_2NH_2$

This is phenylamine, an aromatic amine with an aryl group attached to the amino group.

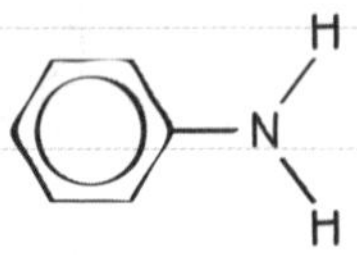

Amines from halogenoalkanes

Primary aliphatic amines can be made from halogenoalkanes.

This is done by mixing a halogenoalkane with concentrated ammonia, or by:

 1 heating a halogenoalkane with excess ammonia gas

 2 under pressure in a sealed container.

1-bromopropane reacts with excess ammonia to form propylamine. There are two steps in the reaction mechanism (you do not need to know it):

1. $CH_3CH_2CH_2Br + NH_3 \rightarrow CH_3CH_2CH_2\overset{+}{N}H_3 + Br^-$

acting as a nucleophile

2. $CH_3CH_2CH_2\overset{+}{N}H_3 + NH_3 \rightarrow CH_3CH_2CH_2NH_2 + NH_4^+$

acting as a Brønsted−Lowry base

Overall:
$CH_3CH_2CH_2Br + 2NH_3 \rightarrow CH_3CH_2CH_2NH_2 + NH_4Br$

Amines from nitriles

Primary aliphatic amines can be made by reducing nitriles using $LiAlH_4$ in dry ether.

For example, propanenitrile can be reduced to form propylamine: ——the reducing agent

$CH_3CH_2CN + 4[H] \rightarrow CH_3CH_2CH_2NH_2$

Which method should you use?

The N atom in the amine group has a lone pair of electrons, so amines can act as nucleophiles.

This means that further substitution can happen when you use halogenoalkanes, leading to the production of secondary and tertiary amines.

This is why you need **excess** ammonia — it reduces the chance of these compounds forming.

There is only one organic product when primary amines are prepared by the reduction of nitriles.

Worked example

Phenylamine can be produced by the reduction of nitrobenzene: $C_6H_5NO_2 + 6[H] \rightarrow C_6H_5NH_2 + 2H_2O$
State the reducing agent used and describe the reaction conditions needed. **(3 marks)**

The reducing agent is tin mixed with concentrated hydrochloric acid. The reaction mixture must be heated under reflux.

Phenylamine is basic (see page 133) so it will react with the acid to form the phenylammonium ion.

This is converted to phenylamine by adding sodium hydroxide solution:

$C_6H_5NH_3^+ + OH^- \rightarrow C_6H_5NH_2 + H_2O$

Now try this

1 (a) Name a bromoalkane and a nitrile that could be used to produce butylamine. **(2 marks)**
  (b) Write an equation to show the reaction using each reactant above. **(2 marks)**
  (c) Explain why a secondary amine, dibutylamine, would form if the bromoalkane and ammonia were used in a 1 : 1 ratio. **(3 marks)**
2 Suggest why phenylamine is not produced by reacting benzene with ammonia. **(2 marks)**

Ammonia is a nucleophile. What type of reactions does benzene normally undergo?

Amines as bases

Amines can act as Brønsted–Lowry bases, and react with water and with acids.

Amines as bases

Primary aliphatic amines are soluble in water because they can form **hydrogen bonds** with water molecules. They also react with water.

The nitrogen atom in an amine molecule (such as butylamine) has a lone pair of electrons.

$$H-\underset{\underset{H}{|}}{\overset{\overset{H}{|}}{C}}-\underset{\underset{H}{|}}{\overset{\overset{H}{|}}{C}}-\underset{\underset{H}{|}}{\overset{\overset{H}{|}}{C}}-\underset{\underset{H}{|}}{\overset{\overset{H}{|}}{C}}-\overset{\overset{H}{\diagup}}{\underset{\underset{H}{\diagdown}}{N:}}$$

This means:
• Amines can act as **Brønsted–Lowry bases**.

They form alkaline solutions in water. For example:

$CH_3CH_2CH_2CH_2NH_2(aq) + H_2O(l)$
$\rightleftharpoons CH_2CH_2CH_2CH_2NH_3^+(aq) + OH^-(aq)$

Ammonia as a base

Ammonia, NH_3, is soluble in water because it can form hydrogen bonds with water molecules.

It also reacts with water.

The nitrogen atom in an ammonia molecule has a lone pair of electrons.

$$H-N:\begin{smallmatrix}H\\\diagup\\\diagdown\\H\end{smallmatrix}$$

This means:
• Ammonia can act as a Brønsted–Lowry base.

It dissolves in water to form an alkaline solution:

$$NH_3(aq) + H_2O(l) \rightleftharpoons NH_4^+(aq) + OH^-(aq)$$

Reaction does not go to completion

Ammonia is only partially dissociated in solution, so it is a weak base. Its pK_a at 298 K is 9.25 (see page 81) for more details about pK_a values).

Basicity of primary amines

Compared to ammonia, primary aliphatic amines:

1 are stronger bases

2 have higher pK_a values (and lower K_a values)

3 form solutions with higher pH values (when they are at the same concentration).

As the alkyl chain length increases, the base strength increases (although the differences become smaller).

Substance	Formula	pK_a
Ammonia	NH_3	9.25
Methylamine	CH_3NH_2	10.64
Ethylamine	$CH_3CH_2NH_2$	10.73

Alkyl groups are **electron-releasing**.

They increase the electron density on the N atom.

The lone pair of electrons becomes more available to form a dative bond with water or an H^+ ion.

Basicity of aromatic amines

Compared to ammonia, primary aromatic amines:

1 are weaker bases

2 have lower pK_a values (and higher K_a values)

3 form solutions with lower pH values (when they are at the same concentration).

The pK_a of phenylamine (below) is 4.62 at 298 K.

The lone pair of electrons on the nitrogen atom delocalises with the delocalised π electrons in the benzene ring.

This decreases the electron density on the N atom.

The lone pair of electrons becomes less available to form a dative bond with water or an H^+ ion.

Worked example

Butylamine reacts with dilute hydrochloric acid. Write an equation to show the reaction. State the type of product formed and name it. **(2 marks)**

$CH_3CH_2CH_2CH_2NH_2 + HCl$
$\rightarrow CH_3CH_3CH_3CH_2NH_3Cl$

The product is a salt, butylammonium chloride.

Now try this

1 Predict the pK_a of propylamine and explain your answer. **(3 marks)**

2 Suggest why phenylmethylamine (right) is a stronger base than ammonia, even though it contains an aryl group. **(3 marks)**

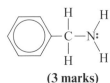

Other reactions of amines

Primary aliphatic amines can react with copper(II) ions to form complex ions, and with halogenoalkanes.

Amines and copper(II) ions

Amines have a lone pair of electrons on their nitrogen atom, just as ammonia does.

Primary aliphatic amines can take part in reactions with copper(II) ions in solution.

For example, with copper(II) ions, butylamine:

 forms a blue precipitate:

$$[Cu(H_2O)_6]^{2+}(aq) +$$
$$2CH_3CH_2CH_2CH_2NH_2(aq)$$
$$\downarrow$$
$$[Cu(H_2O)_4(OH)_2](s) + 2CH_3CH_2CH_2CH_2NH_4^+(aq)$$

An **acid–base** reaction.

 forms a deep blue solution when in excess:

$$[Cu(H_2O)_4(OH)_2](s) + 4CH_3CH_2CH_2CH_2NH_2(aq)$$
$$\downarrow$$
$$[Cu(CH_3CH_2CH_2CH_2NH_2)_4(H_2O)_2]^{2+}(aq) + 2H_2O(l)$$
$$+ 2OH^-(aq)$$

A **ligand-exchange** reaction.

Ammonia and copper(II) ions

Ammonia reacts with copper(II) ions in solution:

pale blue solution   blue precipitate

$$[Cu(H_2O)_6]^{2+}(aq) \rightarrow [Cu(H_2O)_4(OH)_2](s)$$
$$+ 2NH_3(aq) \qquad\qquad + 2NH_4^+(aq)$$

The precipitate redissolves in excess ammonia:

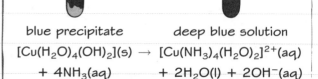

blue precipitate   deep blue solution

$$[Cu(H_2O)_4(OH)_2](s) \rightarrow [Cu(NH_3)_4(H_2O)_2]^{2+}(aq)$$
$$+ 4NH_3(aq) \qquad\qquad + 2H_2O(l) + 2OH^-(aq)$$

Amines and halogenoalkanes

Halogenoalkanes undergo **nucleophilic substitution** reactions with ammonia to form amines (see page 132 for more details). Halogenoalkanes can undergo the same type of reactions with primary aliphatic amines to form secondary amines. In general:

$$R_1NH_2 + R_2X \rightarrow R_1NHR_2 + HX \text{——— X is a halogen atom}$$

amine alkyl group   halogenoalkane alkyl group

For example, butylamine reacts with chloroethane to form *N*-ethylbutylamine and hydrogen chloride:

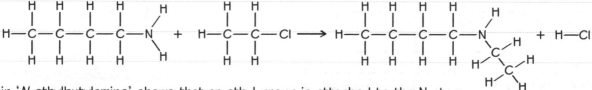

The *N* in '*N*-ethylbutylamine' shows that an ethyl group is attached to the N atom.

Worked example

Ethylamine reacts with chloroethane to form diethylamine, a secondary amine. Explain why further substitution can happen to produce triethylamine, a tertiary amine. **(2 marks)**

There is a lone pair of electrons on the nitrogen atom of diethylamine. This means that diethylamine can act as a nucleophile, allowing further substitution with chloroethane.

The alkyl groups in secondary and tertiary amines make the lone pair of electrons more available for bonding. The chance of further substitution can be decreased by using the primary amine in excess.

Now try this

Write equations to show the reaction between:
(a) Ethylamine and chloroethane   **(1 mark)**
(b) Diethylamine and chloroethane   **(1 mark)**
(c) Triethylamine and chloroethane   **(1 mark)**

Tertiary amines can react with halogenoalkanes.

For example, triethylamine reacts with chloroethane to form a **quaternary ammonium salt** called *N,N,N*-triethylethanaminium chloride. Such substances are useful as fabric softeners.

Making amides

An amide can be prepared from the reaction between an acyl chloride and an amine.

The amide group

The **amide** group consists of:
- a carbonyl group, C=O
- an amino group, $-NH_2$

These are on the same C atom.

They influence each other and amides have some properties not shared with carbonyl compounds or amines.

In **primary amides**, the amide group is located at carbon atom 1. This is 3-methylbutanamide:

Their names end in -amide and you show the functional group as $CONH_2$ in structural formulae.

Name	Structural formula
Methanamide	$HCONH_2$
Ethanamide	CH_3CONH_2
Propanamide	$CH_3CH_2CONH_2$

Secondary and tertiary amides

In **secondary amides**, one of the H atoms in the amide group is replaced by an alkyl or aryl group.

For example, this is N-methylpropanamide:

methyl group attached to N atom

In **tertiary amides**, both H atoms in the amide group are replaced by alkyl or aryl groups.

For example, this is N-ethyl-N-methylpropanamide:

ethyl and methyl groups attached

Acyl chlorides and ammonia

Acyl chlorides react with concentrated ammonia to produce amides and hydrogen chloride.

For example:

$CH_3COCl + NH_3 \rightarrow CH_3CONH_2 + HCl$
ethanoyl   ethanamide
chloride

This type of reaction is **addition–elimination**.

Acyl chlorides and amines

Acyl chlorides react with primary amines to produce secondary amides and hydrogen chloride. For example, ethanoyl chloride reacts with ethanamide to produce N-ethylethanamide:

$CH_3COCl + CH_3CONH_2 \rightarrow CH_3CONHCH_2CH_3 + HCl$

You do not need to know the reaction mechanism.

Worked example

Ethanoyl chloride reacts with butylamine.
(a) Write an equation for the reaction. **(2 marks)**

$CH_3COCl + CH_3CH_2CH_2CH_2NH_2$
$\rightarrow CH_3CONHCH_2CH_2CH_2CH_3 + HCl$

(b) Name the organic product of the reaction. **(1 mark)**

N-butylethanamide.

Ethanoyl chloride and butylamine:

N-butylethanamide:

Now try this

For each of the following pairs of reactants, write the equation and name the organic product.
(a) Propanoyl chloride and concentrated ammonia **(2 marks)**
(b) Propanoyl chloride and propylamine **(2 marks)**

Making polyamides

Polyamides are condensation polymers containing very many amide functional groups.

What are polyamides?

Polyamides are **condensation polymers**.

They are relatively large molecules formed in **condensation reactions** between many smaller **monomer** molecules.

Page 55 compares condensation polymerisation with addition polymerisation.

Nylon and Kevlar® are common polyamides.

In general, polyamides are made from:

1 a diamine (a compound with two amino groups)

2 a dicarboxylic acid (a compound with two carboxyl groups) or a dioyl dichloride (a compound with two acyl chloride groups).

Nylon 4,4

Butane-1,4-dioic acid is a monomer for nylon 4,4.

two carboxyl groups

two CH₂ groups

The other monomer for nylon 4,4 is butane-1,4-diamine.

four CH₂ groups

two amino groups

A carboxyl group on the dicarboxylic acid can react with an amino group on the diamine.

peptide link

Notice that the organic product has a carboxyl group and an amino group.

Further reactions can happen, leading to a long polymer molecule.

Worked example

The diagram shows the repeat unit for Kevlar®.

Name and draw the structures of the dioyl dichloride and diamine that produce this polymer. **(4 marks)**

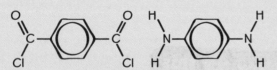

Benzene-1,4-dicarbonyl dichloride and benzene-1,4-diamine.

> The amide link is CONH in the middle of the unit. It is identical to the **peptide link** in polypeptides and proteins (page 138 has more detail about this).

The inorganic product is HCl, not H₂O, if a dioyl dichloride is used instead of a dicarboxylic acid.

Now try this

The structural formula for octane-1,8-dioic acid is $HOOC(CH_2)_6COOH$, and the structural formula for hexane-1,6-diamine is $H_2N(CH_2)_6NH_2$. They can react together to form nylon 6,8.

(a) Draw the displayed formula for each monomer. **(2 marks)**

(b) Draw the repeat unit for nylon 6,8. **(1 mark)**

(c) Suggest why the polymer is called nylon 6,8. **(1 mark)**

Amino acids

Amino acids have amphoteric properties and also often show optical activity.

Identifying amino acids

Amino acid molecules have:

- an amino group, $-NH_2$
- a carboxyl group, $-COOH$

These groups are on different carbon atoms so amino acids retain their properties. In general:

one of several possible 'side groups'

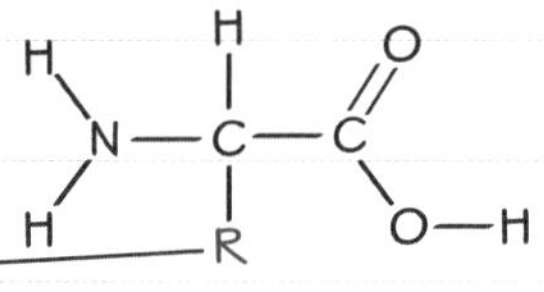

Amino acids found in proteins are 2-amino acids:

 The carboxyl group is located on carbon atom 1.

 The amino group is located on carbon atom 2.

Naming amino acids

Amino acids have the common names used by biologists, and systematic names. For example:

R group	H	CH_3
Structure	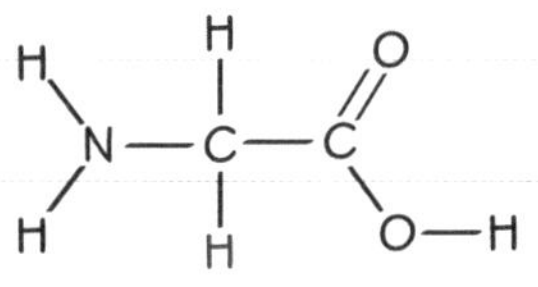	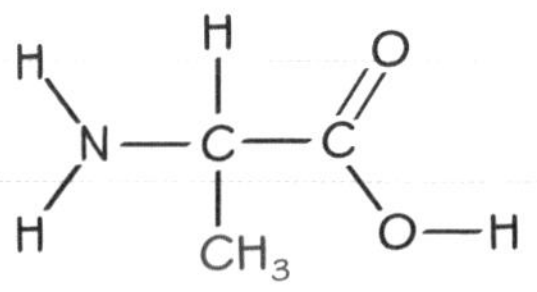
Common name	Glycine	Alanine
Systematic name	Aminoethanoic acid	2-aminopropanoic acid

The locant 2 distinguishes 'alanine' from 3-aminopropanoic acid:

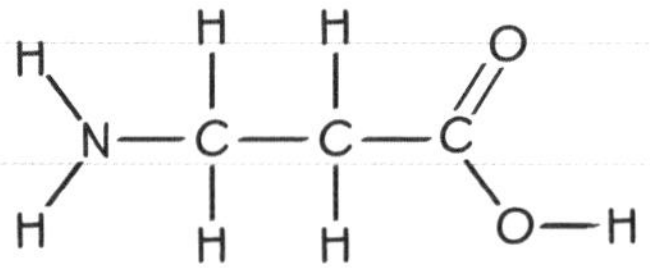

Acid–base properties

Amino acids are **amphoteric**. They can act as acids and as bases and form salts with acids and bases.

Aminoethanoic acid (glycine) acting as an acid:

1. $NH_2CH_2COOH + OH^- \rightarrow NH_2CH_2COO^- + H_2O$
2. $NH_2CH_2COOH + NaOH$
    $\rightarrow NH_2CH_2COO^-Na^+ + H_2O$

Aminoethanoic acid (glycine) acting as a base:

1. $NH_2CH_2COOH + H^+ \rightarrow \overset{+}{N}H_3CH_2COOH$
2. $NH_2CH_2COOH + HCl$
    $\rightarrow [\overset{+}{N}H_3CH_2COOH]Cl^- + H_2O$

Worked example

Explain why glycine (aminoethanoic acid) shows no optical activity but alanine (2-aminopropanoic acid) does. **(2 marks)**

There is no chiral centre in glycine. However, carbon 2 in alanine is asymmetric. It is a chiral centre, so alanine is optically active.

Almost all amino acids are optically active.

They have an **asymmetric carbon atom** with four different groups attached (see page 122).

Naturally occurring amino acids consist of just one **enantiomer** or optical isomer. Amino acids synthesised in the laboratory are **racemic mixtures** with no optical activity.

Zwitterions

Amino acids form **zwitterions** in solution, in which both functional groups are charged.

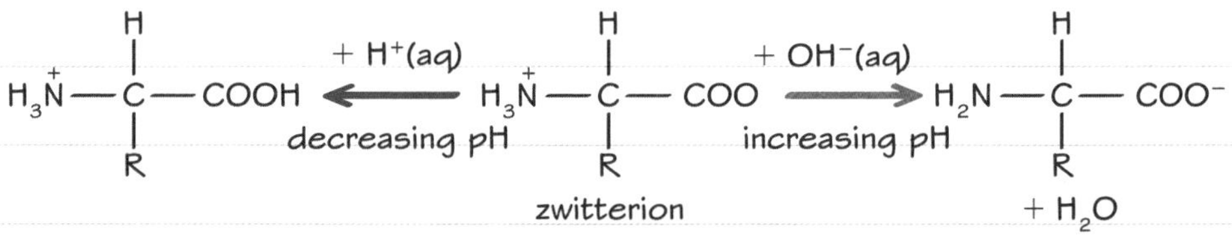

$$\underset{R}{\overset{H}{\underset{|}{\overset{|}{H_3\overset{+}{N}-C-COOH}}}} \xleftarrow[\text{decreasing pH}]{+ H^+(aq)} \underset{\underset{\text{zwitterion}}{R}}{\overset{H}{\underset{|}{\overset{|}{H_3\overset{+}{N}-C-COO^-}}}} \xrightarrow[\text{increasing pH}]{+ OH^-(aq)} \underset{\underset{+ H_2O}{R}}{\overset{H}{\underset{|}{\overset{|}{H_2N-C-COO^-}}}}$$

Now try this

The hydroxyl group in the side chain is referred to as 'hydroxy'.

1 Draw the structures for alanine (2-aminopropanoic acid) in alkaline solution and in acidic solution. **(2 marks)**

2 The displayed structure of serine is shown on the right. State its systematic name. **(1 mark)**

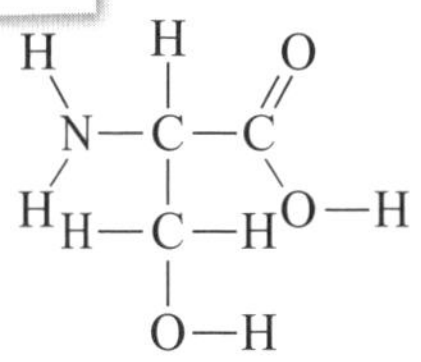

137

Proteins

Amino acids combine to form polypeptides and proteins by condensation polymerisation.

Dipeptides

Two amino acid molecules can react together in condensation reactions between:
- the $-OH$ group in the carboxyl group of one amino acid, and
- one of the H atoms in the amino group of the other amino acid.

A **dipeptide** and water are formed.

For example, aminoethanoic acid (glycine) and 2-aminopropanoic acid (alanine) can react together. The diagram on the right shows how the glycine–alanine dipeptide forms from glycine and alanine.

If they react the other way round, the alanine–glycine dipeptide forms instead (below).

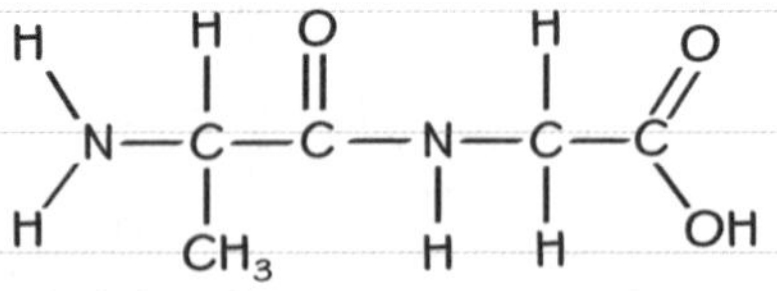

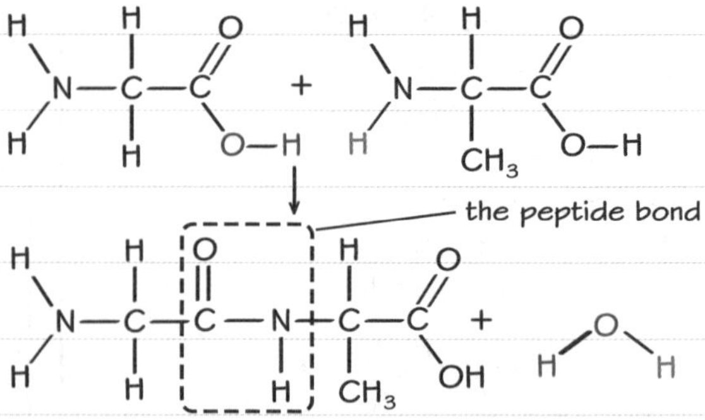

The peptide bond:
- is written as CONH in structural formulae
- is identical to the **amide bond** (see page 136)
- can form at either end of an amino acid molecule.

Polypeptides and proteins

You will see that dipeptides have an amino group at one end and a carboxyl group at the other.

This is the same as for individual amino acids.

As a result, amino acids can continue to join to either end, forming **condensation polymers**.

The amino acid units are called **residues**:
- **Polypeptides** are polymers with fewer than about 20 amino acid residues.
- **Proteins** are polymers with more amino acid residues (they may contain thousands).

Repeat units

You can draw repeat units for simple polypeptides made from one type of amino acid.

This is the repeat unit for poly(2-aminopropanoic acid) or poly(alanine):

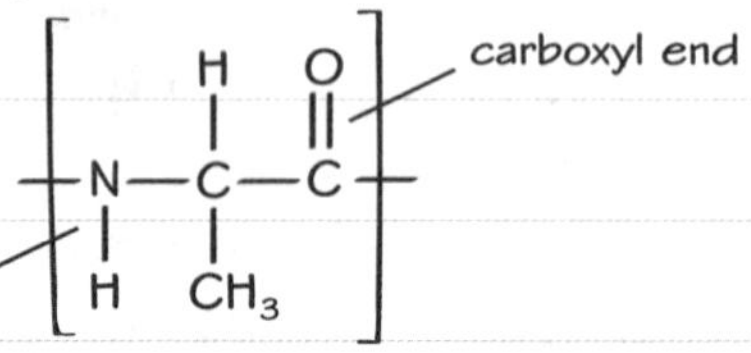

Worked example

Proteins may be hydrolysed in the first step towards analysing their structure.

(a) Describe how a protein may be hydrolysed and identify the type of products formed. **(3 marks)**

Add the protein to concentrated hydrochloric acid and heat for several hours. The products are protonated amino acids.

(b) State a method that can be used to separate the hydrolysis products. **(1 mark)**

Chromatography.

Amino acids can be separated using **paper chromatography** but other more advanced methods are also available (see page 150).

$6\,mol\,dm^{-3}$ hydrochloric acid is suitable and the mixture may need to be heated for 24 hours.

For example, for the glycine–alanine dipeptide described at the top right of this page:

$$H_2NCH_2CONHCH(CH_3)COOH + H_2O + 2H^+$$
$$\downarrow$$
$$H_3\overset{+}{N}CH_2COOH + H_3\overset{+}{N}CH(CH_3)COOH$$

This is the reverse of the process that formed the dipeptide, except that the amino acids are protonated in the acidic conditions.

Now try this

The side group R in the amino acid called cysteine is $-CH_2-SH$. Draw the displayed formulae of the two dipeptides that can be formed by the reaction of glycine with cysteine. **(2 marks)**

Look back at page 137 for the general structure of amino acids.

Exam skills 12

This exam-style question uses knowledge and skills you have already revised. Have a look at pages 132, 133, 137 and 138 for a reminder about amines, amino acids and peptides.

Worked example

(a) Propylamine, $CH_3CH_2CH_3NH_2$, can be prepared from bromoethane, CH_3CH_2Br, using a two-step synthesis. Write an equation for each step involved and name the organic product of Step 1. **(3 marks)**

Step 1 – $CH_3CH_2Br + KCN \rightarrow CH_3CH_2CN + KBr$

Step 2 – $CH_3CH_2CN + 4[H] \rightarrow CH_3CH_2CH_2NH_2$

The organic product in Step 1 is propanenitrile.

(b) Propylamine dissolves in water to form an alkaline solution.

  (i) State the feature of propylamine that allows it to act as a base. **(1 mark)**

It has a lone pair of electrons on its nitrogen atom.

  (ii) Explain which is the stronger base, propylamine or ammonia. **(2 marks)**

Propylamine is the stronger base because its propyl group is electron-releasing. This increases the availability of the lone pair of electrons.

  (iii) Write an equation for the reaction that occurs when propylamine is dissolved in water and explain the role of water in this reaction. **(3 marks)**

$CH_3CH_2CH_3NH_2 + H_2O \rightleftharpoons CH_3CH_2CH_3NH_3^+ + OH^-$

Water is acting as a Brønsted–Lowry acid because it donates a proton to propylamine.

(c) The diagram shows the displayed formula of the amino acid known as aspartic acid.

  (i) State the systematic name for this compound. **(1 mark)**

2-aminobutanedioic acid.

  (ii) Draw the structure of a dipeptide formed by two aspartic acid molecules. **(2 marks)**

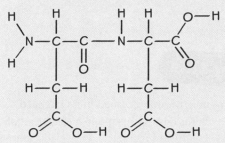

Practical skills   Make sure you can recall the interconversions between different organic compounds.

This will allow you to work out how to produce a particular product, starting with a given starting organic compound.

Remember that the C atom in the nitrile group is counted for naming purposes.

Command word: State

If a question asks you to **state** something, it means you need to recall one or more pieces of information.

The atom with the lone pair of electrons is identified to make the answer clear.

Alkyl groups such as $-CH_3$, $-CH_2CH_3$ and $-CH_2CH_2CH_3$ are electron donating and increase the electron density on the nitrogen atom. Take care: this does not increase the electronegativity of the nitrogen atom.

Make sure you are prepared for synoptic questions, which are drawn from more than one topic.

State symbols were not asked for in this question. You could have used an arrow here instead of the reversible symbol.

Practical skills   Aspartic acid has a chiral carbon atom.

The biologically produced amino acid shows optical activity in solution.

You could determine its effect on plane-polarised light using a polarimeter.

Aspartic acid has one amino group $-NH_2$ but, unusually, two carboxyl groups $-COOH$. When you draw the dipeptide, involve the carboxyl group with the locant of 1 (not the one at the bottom).

Check that the N atom in the peptide bond only has three bonds and all carbon atoms have four bonds.

Grignard reagents

The length of a carbon chain in a molecule can be increased using reactions with Grignard reagents.

What are Grignard reagents?

Grignard reagents are **organometallic** compounds with a carbon–magnesium bond.

In general: $RMgX$ —— halogen, e.g. Br or I
alkyl or aryl group

Grignard reagents are made by:

✓ heating a halogenoalkane under reflux

✓ with magnesium

✓ in a solvent of dry ether.

The dry ether provides anhydrous conditions to prevent the Grignard reagent reacting with water.

Desired product	Grignard reagent with
Carboxylic acid	carbon dioxide
Primary alcohol	methanal
Secondary alcohol	aldehyde (not methanal)
Tertiary alcohol	ketone

Making a carboxylic acid

For example, to make propanoic acid, you need:
- bromoethane to provide the CH_3CH_2 part
- carbon dioxide to provide the COOH part.

1 Form Grignard reagent
$CH_3CH_2Br + Mg \rightarrow CH_3CH_2MgBr$

2 React with carbon dioxide (below 0°C)
$CH_3CH_2MgBr + CO_2 \rightarrow CH_3CH_2COOMgBr$

3 Hydrolyse using dilute acid
$CH_3CH_2COOMgBr + H_2O \xrightarrow{H^+} CH_3CH_2COOH + Mg(OH)Br$

Notice that the number of carbon atoms has increased from 2 (in bromoethane) to 3 (in propanoic acid).

These reactions increase the carbon chain length.

Reactions with carbonyl compounds

Grignard reagents react with carbonyl compounds following this general scheme.

R_1, R_2 and R_3 represent alkyl groups, which can be the same as each other or different from one another.

$$R_1{-}MgBr + \begin{array}{c}R_3\\ \diagup\\ C{=}O\\ \diagdown\\ R_2\end{array} \longrightarrow R_1{-}\underset{R_2}{\overset{R_3}{C}}{-}O{-}MgBr \xrightarrow{H^+} R_1{-}\underset{R_2}{\overset{R_3}{C}}{-}O{-}H + Mg(OH)Br$$

The Grignard reagent reacts with the carbonyl compound. | Intermediate is hydrolysed under acidic conditions. | The final products form. The magnesium compound reacts further with the acid.

The product depends on the choice of halogenoalkane for the Grignard reagent, and the carbonyl compound.

Worked example

(a) Name the bromoalkane and carbonyl compound needed to form pentan-2-ol and show the Grignard reagent formed by the bromoalkane. **(3 marks)**

1-bromopropane and ethanal. The Grignard reagent formed is $CH_3CH_2CH_2MgBr$.

(b) Name the organic product produced using bromomethane and propanone in a Grignard synthesis. **(1 mark)**

2-methylpropan-2-ol.

Using the general reaction scheme above:
$R_1 = CH_3$ (from the bromoalkane)
R_2 and $R_3 = CH_3$ (in the carbonyl compound)
2-methylpropan-2-ol is a tertiary alcohol.

Using the general reaction scheme above:
$R_1 = CH_3CH_2CH_2$ (from the bromoalkane)
$R_2 = H$ (if $R_3 = CH_3$) (in the carbonyl compound)
Pentan-2-ol is then shown as:

$$H_3C{-}CH_2{-}CH_2{-}\underset{H}{\overset{CH_3}{C}}{-}O{-}H$$

Now try this

Name the organic product formed in a Grignard synthesis involving 1-bromopropane and methanal, and write its structural formula. **(2 marks)**

Methods in organic chemistry 1

Organic solids can be purified using recrystallisation, and their purity determined using their melting temperature.

Control measures

When you carry out experiments:

- A **hazard** is a substance or procedure that could cause harm or damage.
- A **risk** is the likelihood that a hazard will actually lead to harm or damage.
- A **control measure** is something you do to reduce the risk presented by a hazard.

When you identify the control measures needed, these should be specific to the experiment.

For example, hydrochloric acid is corrosive at high concentrations. There is a risk of it damaging your skin unless you wear gloves.

Hazard symbols

harmful   corrosive   toxic   flammable   oxidising

explosive   serious longer term health hazard   harmful to the environment   contains gas under pressure

Recrystallisation

If you have prepared an organic solid, you can use **recrystallisation** to remove impurities from it.

You need to choose a solvent in which:

- most or all of the impurities are soluble
- your solid is insoluble or sparingly soluble.

The main steps are:

1. The solid is dissolved in hot solvent using just enough solvent to do it — the minimum volume.

2. The mixture is filtered while it is still hot.

3. The filtrate is cooled in an ice bath.

4. The crystals are separated by filtration using a Buchner funnel, washed on the filter paper using cold solvent, then dried.

If you use too much solvent at Step 1, crystals may not form at Step 3.

Melting point determination

Different organic solids have different melting temperatures. You can estimate purity by measuring the melting temperature and comparing this with the known value (data book).

Your product will be pure if:

- its melting temperature is sharp, rather than spread over several °C
- its melting temperature matches the known value.

The main steps are:

1. Add some powdered solid to a melting point capillary tube (sealed at one end).

2. Place the capillary tube in the melting point apparatus with a thermometer.

3. Raise the temperature slowly to obtain an accurate melting temperature.

Worked example

An organic solid is purified by recrystallisation. Which impurities are removed by the hot filtration and which by the cold filtration? **(2 marks)**

Insoluble impurities are removed by the hot filtration and soluble impurities are removed by the cold filtration.

Practical skills — Drying

Purified organic solids are dried after recrystallisation, by leaving them in:

- a warm place, such as a warm oven
- a desiccator with a drying agent.

Organic liquids can be dried too. You can find details about this on page 61.

Now try this

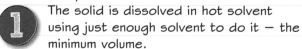
Think about how the solubility of most solids varies with temperature.

1 Explain why crystals of the desired organic product form during recrystallisation. **(2 marks)**

2 Ethanol can be oxidised to ethanal by heating it with potassium dichromate(VI) solution, acidified with dilute sulfuric acid, under distillation conditions. Identify two hazards in this synthesis and suggest a suitable control measure for each one. **(4 marks)**

Methods in organic chemistry 2

> **Practical skills**   Refluxing and simple distillation are useful for preparing organic liquids, while fractional distillation and steam distillation are useful for purifying them.

Refluxing

Refluxing allows you to heat a reaction mixture for a long time without losing any liquid.

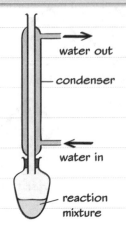

You set up the condenser in a different way from when you use it for distillation (see right).

Fractional distillation

You can use **fractional distillation** to separate more than one liquid from a mixture of liquids.

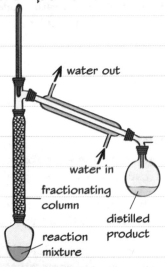

Simple distillation

Simple distillation allows the product to leave the reaction mixture as it forms.

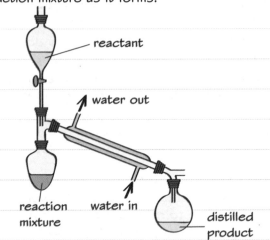

Steam distillation

You can use **steam distillation** to separate an insoluble liquid from an aqueous solution.

It involves:

- passing steam into the reaction mixture
- the steam bubbling through the mixture brings both liquids to the surface
- both liquids can form part of the liquid that evaporates.

The insoluble liquid is removed from the reaction mixture below its boiling temperature, reducing the chance of it decomposing. For example, the boiling temperature of phenylamine is 184 °C but a mixture of phenylamine and water distils at 98 °C.

Worked example

Describe how you could use boiling temperature data to determine the purity of an organic liquid.   **(2 marks)**

Compare the boiling temperature of the organic liquid with its known value (from a data book). The closer the two temperatures, the purer the liquid is.

You could use simple distillation apparatus set up with a thermometer instead of a dropping funnel.

You may reach an incorrect conclusion about the identity and purity of the liquid if, for example:

- your thermometer is incorrectly calibrated
- the liquid shares the same boiling temperature with another liquid.

Solvent extraction

You can use a **separating funnel** to separate two **immiscible** liquids (liquids that do not mix).

This method works because the liquids form two layers, one above the other.

You can find details about this on page 61.

Now try this

Which of the techniques described on this page:

(a) involves continuous evaporation and condensation?   **(1 mark)**

(b) is suitable for separating limonene, an insoluble liquid that boils at 176 °C, from water?   **(1 mark)**

Reaction pathways

You need to be able to plan reaction schemes with up to four steps and involving organic compounds.

Flow chart of some aliphatic reaction pathways

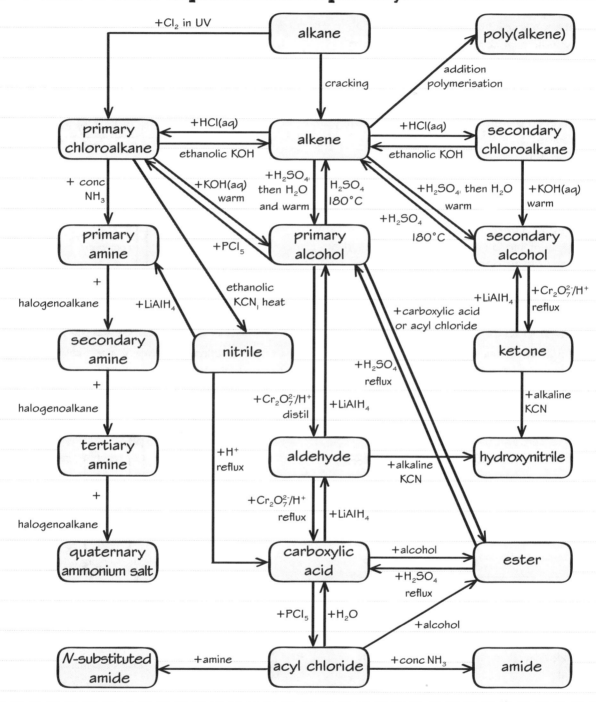

Outline how you could produce 1,4-diaminobutane, $H_2N(CH_2)_4NH_2$, starting with ethene, C_2H_4.

(3 marks)

React ethene with Br_2, forming $BrCH_2CH_2Br$. Warm this with ethanolic KCN to form $NCCH_2CH_2CN$. Reduce this product to $H_2N(CH_2)_4NH_2$ using $LiAlH_4$.

Make a flow chart for the reactions of benzene including halogenation, nitration and Friedel–Crafts acylation and alkylation.   **(6 marks)**

Look at pages 128–132 for help.

Chromatography

Chromatography separates components of a mixture between a mobile phase and a stationary phase.

Two phases

Chromatography is used to:
- separate a mixture into its components
- identify individual components of a mixture.

There are different forms of chromatography.

They all have:

 A **stationary phase** which does not move, such as paper, silica, alumina or resin beads

 A **mobile phase** which moves through the stationary phase, usually a solvent or gas.

Each component in a mixture is attracted to the two phases by different amounts:
- A component more strongly attracted to the stationary phase does not travel very far.
- A component more strongly attracted to the mobile phase travels further in the same time.

R_f values in one-way chromatography

The stationary phase in **paper chromatography** is a sheet or strip of absorbent paper.

In **thin layer chromatography** or 'TLC' it is silica or alumina coated onto glass or plastic.

The mobile phase in both cases is a liquid solvent.

You calculate the R_f value of a component using:

$$R_f = \frac{\text{distance travelled by component}}{\text{distance travelled by solvent}}$$

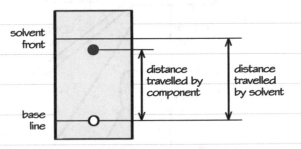

Retention times in HPLC and GC

In HPLC (high performance liquid chromatography) the mobile phase, a solvent, is pumped through the column at high pressure.

In GC (**gas chromatography**) the mobile phase, an inert carrier gas such as nitrogen, pushes the components through the column.

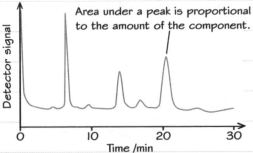

The **retention time** is the time between the instant the sample is injected and the peak of its detection. It is different for different components.

Column chromatography

In column chromatography, the stationary phase is packed into a tube or 'column'.

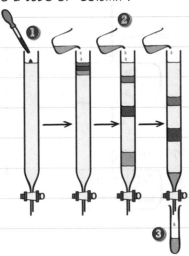

1. Mixture applied to the top of the column.
2. Solvent added – components move past the stationary phase.
3. Different components leave at the bottom.

The column can be a simple glass tube or a burette. In HPLC and GC it is a metal tube.

Worked example

GC may be used in conjunction with mass spectrometry. State a suitable application for this method and explain why it may be done. **(3 marks)**

This can be used in forensic science.
GC separates the components of a sample.
Although the amount of each component is very small, the mass spectrum reveals information about its structure.

Now try this

Calculate the R_f value for the component in the paper chromatogram on this page. **(1 mark)**

Measure the two distances using a ruler.

Functional group analysis

Information about the structure of an organic compound may be obtained using simple laboratory tests.

Halogenoalkanes

The functional group in a **halogenoalkane** is a halogen atom. You can determine the identity of the halogen atom in a chloroalkane, bromoalkane or iodoalkane using silver nitrate solution:

1. Add a mixture of ethanol and sodium hydroxide solution to a sample of the halogenoalkane.

2. Warm the mixture.

3. Acidify with dilute nitric acid.

4. Add silver nitrate solution and observe the colour of the silver halide precipitate formed.

5. Carry out a confirmatory test with ammonia solution to be sure of the result.

Steps 1 and 2 cause a **substitution** reaction in which an alcohol forms and a halide ion is released.

Step 3 stops hydroxide ions reacting with silver ions, which would give a brown precipitate.

Halogen	chlorine	bromine	iodine
Halide ion released	chloride	bromide	iodide
Colour of precipitate	white	cream	yellow
Test with dilute NH_3	ppt redissolves	no change	no change
Test with conc. NH_3	ppt redissolves	ppt redissolves	no change

Aldehydes and ketones

The functional group in **aldehydes** and **ketones** is the carbonyl group, $C=O$.

They both react with 2,4-DNPH (Brady's reagent) to form orange–yellow precipitates, but there are several different tests you can use to distinguish between these carbonyl compounds.

	Aldehyde	Ketone
Fehling's or Benedict's solution	turns orange or red precipitate forms	no change (stays blue)
Tollens' reagent	silver mirror or grey precipitate forms	no change (stays colourless)
$K_2Cr_2O_7$ acidified with dilute H_2SO_4	turns green	no change (stays orange)

Ethanal, and certain ketones, react with alkaline solutions of iodine to form a yellow precipitate of triiodomethane, CHI_3 ('iodoform' – see page 124 for more details about these reactions).

Carboxylic acids

Carboxylic acids contain the carboxyl group, $-COOH$ and show acidic properties. Carboxylic acids:

- turn universal indicator solution red or orange

- react with magnesium and other reactive metals to produce hydrogen gas

- react with hydrogencarbonates and carbonates, such as $Na_2CO_3(aq)$, to produce bubbles of CO_2

- react with phosphorus(V) chloride, PCl_5, to produce misty fumes of hydrogen chloride, HCl.

Worked example

Describe a laboratory test to distinguish between cyclohexane and cyclohexene. **(2 marks)**

Add a few drops of bromine water to each substance in a test tube, stopper the tube, then shake to mix. With cyclohexane, the orange-brown colour of the lower, aqueous layer will remain, but with cyclohexene it will be decolourised.

Bromine could be used but is more hazardous.

Acidified potassium manganate(VII) oxidises the C=C bond and changes from purple to colourless.

However, it is less specific as a test for unsaturation.

Now try this

You are given three bottles, each containing a primary, secondary or tertiary alcohol. Describe simple laboratory tests to distinguish between them. Describe what you would do and what you would expect to see for each type of alcohol. **(6 marks)**

Primary and secondary alcohols can be oxidised to form carbonyl compounds, but tertiary alcohols are resistant to oxidation.

Combustion analysis

Combustion analysis can be used to determine the masses of the elements in a compound and from this information to deduce its empirical formula.

Data from combustion analysis

The data obtained from a **combustion analysis** experiment include the mass of:

- the sample of the compound
- carbon dioxide produced
- water produced.

From these data, you can calculate the mass of:

- carbon, C, in the sample
- hydrogen, H, in the sample.

If the compound contains oxygen, you can also calculate the mass of oxygen it contained:

$$\text{mass of O} = \text{mass of sample} - (\text{mass of C} + \text{mass of H})$$

Apparatus used

In combustion analysis:

1. The mass of the sample is measured.
2. The sample is heated in oxygen.
3. The increase in mass of each absorber during the experiment is measured.

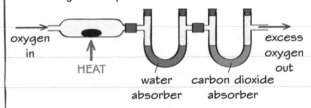

oxygen in   HEAT   water absorber   carbon dioxide absorber   excess oxygen out

---

Worked example

Compound X consists of carbon, hydrogen and oxygen only. A sample was completely burnt in oxygen. The table shows the results of the experiment.

Substance	Mass/g
Compound X	1.76
Carbon dioxide	3.52
Water	1.44

(a) Calculate the mass of carbon present in compound X. **(1 mark)**

$$\text{mass of C} = \text{mass of CO}_2 \times \frac{A_r \text{ of C}}{M_r \text{ of CO}_2}$$

$$\text{mass of C} = 3.52 \times \frac{12.0}{44.0} = 0.96\,g$$

(b) Calculate the mass of hydrogen present in compound X. **(1 mark)**

$$\text{mass of H} = \text{mass of H}_2O \times \frac{2 \times A_r \text{ of C}}{M_r \text{ of H}_2O}$$

$$\text{mass of H} = 1.44 \times \frac{2 \times 1.0}{18.0} = 0.16\,g$$

> Remember that the formula for water is H_2O.
>
> Each water molecule contains two H atoms.
>
> This is where the factor of 2 comes from in the equation above.

(c) Calculate the mass of oxygen present in compound X. **(1 mark)**

Mass of O $= 1.76 - (0.96 + 0.16)\,g$
$\qquad = 0.64\,g$

(d) Calculate the empirical formula of X. **(3 marks)**

C   H   O

$\frac{0.96}{12.0} = 0.080$ $\frac{0.16}{1.0} = 0.16$ $\frac{0.64}{16.0} = 0.040$

$0.080 \div 0.040$ $0.16 \div 0.040$ $0.040 \div 0.040$
$\quad = 2 \qquad\qquad = 4 \qquad\qquad = 1$

Empirical formula is C_2H_4O.

> You can determine the molecular formula of compound X if you know its relative formula mass.
>
> You can find more detailed information about these calculations on page 34.
>
> You cannot deduce the structural formula and identity of compound X from these data alone — you need additional data from functional group analysis (page 145) or instrumental analysis.

Now try this

Compound Y consists of carbon, hydrogen and oxygen only. The table shows the results of a combustion analysis experiment. Determine the empirical formula of compound Y, showing your working out clearly. **(6 marks)**

Substance	Mass/g
Compound Y	0.88
Carbon dioxide	1.76
Water	0.72

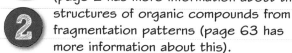

High-resolution mass spectra

High-resolution mass spectrum data can be used to deduce possible molecular formulae of a compound.

High-resolution mass spectrometry

Remember that **mass spectrometry** can be used to determine the:

1 relative atomic masses of elements and relative molecular masses of compounds (page 2 has more information about this)

2 structures of organic compounds from fragmentation patterns (page 63 has more information about this).

High-resolution mass spectrometry produces relative molecular mass values precise to 4 decimal places or more. With these, you can deduce the possible structure of an organic compound.

Some relative atomic masses

Element	Symbol	A_r 1 d.p.	A_r 4 d.p.
Hydrogen	H	1.0	1.0078
Carbon	C	12.0	12.0107
Nitrogen	N	14.0	14.0067
Oxygen	O	16.0	15.9994

Values actually have a range, depending on the source of the sample.

Be careful! The A_r for the isotope ^{12}C is defined as 12.0000 exactly, but a sample of carbon contains other isotopes such as ^{14}C, so its A_r is different from this.

Accurate M_r values

Ethanoic acid and urea have the same low-resolution M_r value: 60.0

Their high-resolution M_r values are different.

If you are given a high-resolution M_r value for an unknown substance (from its **molecular ion** peak in the mass spectrum) you can determine what it is likely to be from a range of structures:
- Calculate the M_r values for each structure.
- Compare the M_r values to the given value.

ethanoic acid, $M_r = 60.0514$

urea, $M_r = 60.0547$

Worked example

An organic compound X is thought to be ethane, C_2H_6, or methanal, HCHO. The relative molecular mass for the compound is 30.0262, determined by high-resolution mass spectrometry.

(a) Calculate the relative molecular masses of these two compounds. **(2 marks)**

M_r of C_2H_6
= (2 × 12.0107) + (6 × 1.0078)
= 24.0214 + 6.0468 = 30.0682

M_r of HCHO
= (2 × 1.0078) + (1 × 12.0107) + (1 × 15.994)
= 2.0156 + 12.0107 + 15.9994 = 30.0257

(b) Explain which of the two substances X is likely to be. **(2 marks)**

X is likely to be methanal because its M_r is closer to the M_r of methanal.

Remember that the Periodic Table in the Data Book shows A_r values precise to 1 decimal place.
You will be given accurate A_r values if you need them for these calculations in the examination.

Working out is shown and all values are expressed to 4 decimal places (not to 4 significant figures).

The two M_r values may not be identical because of differences between the isotopic abundance in the sample and in the substances used to produce the M_r data.

You could carry out a chemical test to confirm your conclusion. For example, methanal forms a silver mirror with Tollens' reagent but ethane does not.

Now try this

An organic compound Y is thought to be propanone, CH_3COCH_3, or ethene-1,2-diamine, $H_2NCH=CHNH_2$. The relative molecular mass for the compound is 58.0795, determined by high-resolution mass spectrometry. Determine the likely identity of Y and explain your answer. **(4 marks)**

¹³C NMR spectroscopy

¹³C NMR spectroscopy provides information about the positions of ¹³C atoms in a molecule.

NMR spectroscopy

NMR stands for **nuclear magnetic resonance**.

Nuclei with an odd mass number have residual **spin**, and can be affected by external magnetic fields.

These include the nuclei of ¹³C and ¹H atoms.

The difference in energy between the two spin states corresponds to radio-frequency electromagnetic radiation.

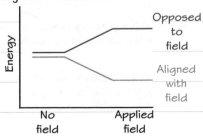

The nuclei of the different atoms in a molecule absorb different frequencies. These **resonance frequencies** provide an **NMR spectrum**.

Interpreting ¹³C NMR spectra

In a ¹³C NMR spectrum:

1 The number of peaks tells you how many different **chemical environments** there are.

2 The chemical shift of each peak gives you information about the chemical environment.

For example, the ¹³C NMR spectrum of 2-methylprop-2-enoic acid shows that there are four chemical environments in the molecule.

TMS and chemical shift

Tetramethylsilane (TMS), $Si(CH_3)_4$, is used as the **reference standard** in NMR spectroscopy:

1 It is unreactive and does not react with most organic compounds.

2 Its ¹³C nuclei are highly **shielded** from the magnetic field because silicon has a low electronegativity, so a sharp peak is produced **upfield** (to the right) of most other ¹³C nuclei.

The position of a peak in an NMR spectrum is given by its **chemical shift**, δ, measured in parts per million or ppm.

The chemical shift of TMS is 0 ppm by definition.

Be careful! Nuclear spin is not the same as electron spin, and nuclear shielding is not the same as electron shielding.

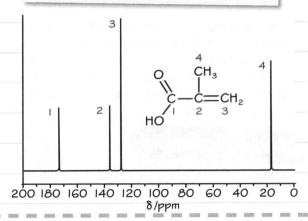

Worked example

Predict the number of peaks that would be present in the ¹³C NMR spectrum of cyclohexene and explain your answer. **(2 marks)**

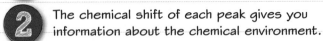
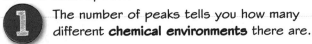

There will be three peaks because there are three chemical environments, identified on the diagram.

The Data Booklet gives chemical shift data for different chemical environments.

Working clockwise:

1. The top right CH is adjacent to =CH−CH₂−CH₂−CH₂−CH₂.

2. The bottom right CH₂ is adjacent to −CH₂−CH₂−CH₂−CH=CH.

3. The bottom CH₂ is adjacent to −CH₂−CH₂−CH=CH−CH₂.

This gives three different environments.

These are mirrored on the other side, where there are equivalent environments.

Now try this

Predict, and explain, the number of peaks in the ¹³C NMR spectra for:
(a) Benzene   **(2 marks)**
(b) 2-methylbutane.   **(2 marks)**

Proton NMR spectroscopy

Proton NMR spectroscopy provides information about the relative numbers of hydrogen atoms in the different chemical environments in a molecule.

Interpreting proton NMR spectra

In a proton NMR spectrum:

1 The number of peaks tells you how many different **chemical environments** there are.

2 The chemical shift of each peak gives you information about the chemical environment.

3 The area under the peak tells you the **relative** numbers of hydrogen atoms in each chemical environment — it may equal the actual number of atoms but not always.

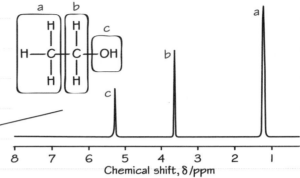

low-resolution spectrum of ethanol

- There are three peaks so three chemical environments.
- Ratio is 3 : 2 : 1 in the relative numbers of hydrogen atoms in each chemical environment.

Solvents in proton NMR

The solvents used in proton NMR spectroscopy must not contain ^1H atoms or they will produce their own peaks in the spectrum.

Suitable solvents include:

☑ tetrachloromethane, CCl_4, which does not contain hydrogen atoms

☑ **deuterated** solvents such as $CDCl_3$ (trichloromethane $CHCl_3$ containing ^2H not ^1H).

TMS in proton NMR

As in ^{13}C NMR spectroscopy, tetramethylsilane (TMS) is the reference standard.

- It is unreactive.
- Its ^1H nuclei are highly shielded.
- It has 12 equivalent hydrogen atoms (in four CH_3 groups) giving a high, sharp peak.

It may help if you draw the displayed formulae and circle each chemical environment.

Worked example

Propan-1-ol and propan-2-ol are two isomers of C_3H_8O. Predict the numbers of peaks that would be present in their low-resolution proton NMR spectra and the peak area ratios. **(2 marks)**

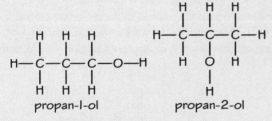

propan-1-ol   propan-2-ol

There will be four peaks in the ratio 3 : 2 : 2 : 1 in the spectrum for propan-1-ol.

There will be three peaks in the ratio 6 : 1 : 1 in the spectrum for propan-2-ol.

Now try this

1 State the number of different chemical environments in:
   (a) Methane, CH_4 **(1 mark)**
   (b) Ethane, C_2H_6 **(1 mark)**
   (c) Propane, C_3H_8. **(1 mark)**

2 (a) Predict the number of peaks in the low-resolution proton NMR spectrum of the compound shown below and explain your answer. **(2 marks)**

$$H_3C-\overset{\overset{\displaystyle CH_3}{|}}{\underset{\underset{\displaystyle CH_3}{|}}{C}}-\overset{\overset{}{}}{\underset{\underset{\displaystyle O}{\|}}{C}}-CH_3$$

   (b) State the number of hydrogen atoms in each chemical environment. **(1 mark)**
   (c) State the simplest peak area ratio for the peaks in the spectrum. **(1 mark)**

The four peaks for propan-1-ol correspond to:
- CH_3, CH_2, CH_2, O—H

The hydrogen atoms in the two CH_2 groups are in different chemical environments. Although there are two CH_3 groups in propan-2-ol, they are in the same chemical environment, so their hydrogen atoms produce a single peak.

The remaining peaks correspond to:
- CH, O—H

Splitting patterns

High-resolution proton NMR spectroscopy provides detailed information for deducing structures.

n + 1 rule

Non-equivalent protons are in different chemical environments. **Peak splitting** happens when there are adjacent non-equivalent hydrogen atoms in a molecule. To work out how a peak splits because of **spin–spin coupling**:

 Count the adjacent hydrogen atoms.

 The peak is split by this number plus one (called the $n + 1$ rule).

A singlet is just one peak. It is:
- produced when there is no spin–spin coupling
- typical of a hydroxyl group, −OH.

Peak splitting produces **multiplets**.

Maths skills — Peak area ratios of multiplets

You can find the peak area ratio of a multiplet using **Pascal's triangle**.

Adjacent H atoms	Multiplet name	Peak area ratio
0	singlet	1
1	doublet	1:1
2	triplet	1:2:1
3	quartet	1:3:3:1
4	quintet	1:4:6:4:1
5	sextet	1:5:10:10:5:1
6	septet	1:6:15:20:15:6:1

Inside the triangle, each number in a row is the sum of the two numbers immediately above it.

Predicting splitting patterns

You can predict the splitting pattern for a compound, for example, ethanol.

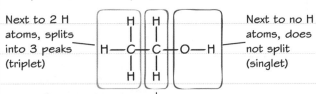

Next to 2 H atoms, splits into 3 peaks (triplet)

Next to no H atoms, does not split (singlet)

Next to 3 H atoms, splits into 4 peaks (quartet)

The **integration trace** represents the peak areas. Its height is proportional to the number of H atoms.

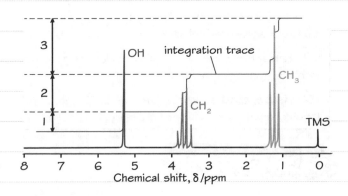

Worked example

The table shows data from the proton NMR spectrum of a compound with the molecular formula $C_4H_8O_2$.

δ/ppm	Splitting pattern	Integration value
1.15	triplet	3
2.32	quartet	2
3.67	singlet	3

(a) State what can be deduced about the structure of the compound from each peak. **(3 marks)**

The triplet is due to a CH_3 group adjacent to a CH_2 group. The quartet is due to a CH_2 group adjacent to a CH_3 group. The singlet is due to a CH_3 group, not adjacent to non-equivalent hydrogen atoms.

(b) Deduce the structure of the compound. **(1 mark)**

$CH_3CH_2COOCH_3$

Now try this

Predict the splitting pattern for 1-chloro-1-hydroxypentan-3-one in high-resolution proton NMR spectroscopy. In your answer, identify the group responsible for each multiplet or singlet. **(5 marks)**

The CH_3 and CH_2 are next to each other: CH_3CH_2 The singlet might have been due to hydrogen atoms in equivalent −OH groups. However, there are 8 H atoms in the formula with 5 in the CH_3CH_2 group: you would need 3 −OH groups but there are only 2 O atoms in the formula.

Periodic table

Key

Atomic (proton number)
Atomic symbol
Name
Relative atomic mass

| | 1 | H | Hydrogen | 1.0 | |

Group	(1)	(2)	(3)	(4)	(5)	(6)	(7)	(8)	(9)	(10)	(11)	(12)	(13)	(14)	(15)	(16)	(17)	(18) 8
Period 1																		2 He Helium 4.0
Period 2	3 Li Lithium 6.9	4 Be Beryllium 9.0											5 B Boron 10.8	6 C Carbon 12.0	7 N Nitrogen 14.0	8 O Oxygen 16.0	9 F Fluorine 19.0	10 Ne Neon 20.2
Period 3	11 Na Sodium 23.0	12 Mg Magnesium 24.3											13 Al Aluminium 27.0	14 Si Silicon 28.1	15 P Phosphorus 31.0	16 S Sulfur 32.1	17 Cl Chlorine 35.5	18 Ar Argon 39.9
Period 4	19 K Potassium 39.1	20 Ca Calcium 40.1	21 Sc Scandium 45.0	22 Ti Titanium 47.9	23 V Vanadium 50.9	24 Cr Chromium 52.0	25 Mn Manganese 54.9	26 Fe Iron 55.8	26 Co Cobalt 58.9	28 Ni Nickel 58.7	29 Cu Copper 63.5	30 Zn Zinc 65.4	31 Ga Gallium 69.7	32 Ge Germanium 72.6	33 As Arsenic 74.9	34 Se Selenium 79.0	35 Br Bromine 79.9	36 Kr Krypton 83.8
Period 5	37 Rb Rubidium 85.5	38 Sr Strontium 87.6	39 Y Yttrium 88.9	40 Zr Zirconium 91.2	41 Nb Niobium 92.9	42 Mo Molybdenum 95.9	43 Tc Technetium (98)	44 Ru Ruthenium 101.1	45 Rh Rhodium 102.9	46 Pd Palladium 106.4	47 Ag Silver 107.9	48 Cd Cadmium 112.4	49 In Indium 114.8	50 Sn Tin 118.7	51 Sb Antimony 121.8	52 Te Tellurium 127.6	53 I Iodine 126.9	54 Xe Xenon 131.3
Period 6	55 Cs Caesium 132.9	56 Ba Barium 137.3	57 La* Lanthanum 138.9	72 Hf Hafnium 178.5	73 Ta Tantalum 180.9	74 W Tungsten 183.8	75 Re Rhenium 186.2	76 Os Osmium 190.2	77 Ir Iridium 192.2	78 Pt Platinum 195.1	79 Au Gold 197.0	80 Hg Mercury 200.6	81 Tl Thallium 204.4	82 Pb Lead 207.2	83 Bi Bismuth 209.0	84 Po Polonium (209)	85 At Astatine (210)	86 Rn Radon (222)
Period 7	87 Fr Francium (223)	88 Ra Radium (226)	89 Ac* Actinium (227)	104 Rf Rutherfordium (261)	105 Db Dubnium (262)	106 Sg Seaborgium (266)	107 Bh Bohrium (264)	108 Hs Hassium (277)	109 Mt Meitnerium (268)	110 Ds Darmstadtium (271)	111 Rg Roentgenium (272)	112 Cn Copernicium 112		114 Fl flerovium		116 Lv livermorium		

58 Ce Cerium 140.1	59 Pr Praseodymium 140.9	60 Nd Neodymium 144.2	61 Pm Promethium 144.9	62 Sm Samarium 150.4	63 Eu Europium 152.0	64 Gd Gadolinium 157.2	65 Tb Terbium 158.9	66 Dy Dysprosium 162.5	67 Ho Holmium 164.9	68 Er Erbium 167.3	69 Tm Thulium 168.9	70 Yb Ytterbium 173.0	71 Lu Lutetium 175.0
90 Th Thorium 232.0	91 Pa Protactinium (231)	92 U Uranium 238.1	93 Np Neptunium (237)	94 Pu Plutonium (242)	95 Am Americium (243)	96 Cm Curium (247)	97 Bk Berkelium (245)	98 Cf Californium (251)	99 Es Einsteinium (254)	100 Fm Fermium (253)	101 Md Mendelevium (256)	102 No Nobelium (254)	103 Lr Lawrencium (257)

Data booklet

Physical constants

Avogadro constant (L)	$6.02 \times 10^{23} \, mol^{-1}$
Elementary charge (e)	$1.60 \times 10^{-19} \, C$
Gas constant (R)	$8.31 \, J \, mol^{-1} \, K^{-1}$
Molar volume of ideal gas: at r.t.p.	$24 \, dm^3 \, mol^{-1}$
Specific heat capacity of water	$4.18 \, J \, g^{-1} \, K^{-1}$
Ionic product of water (K_w)	$1.00 \times 10^{-14} \, mol^2 \, dm^{-6}$

$1 \, dm^3 = 1 \, 000 \, cm^3 = 0.001 \, m^3$

Correlation of infrared absorption wavenumbers with molecular structure

Group	Wavenumber range/cm^{-1}
C—H stretching vibrations	
Alkane	2962–2853
Alkene	3095–3010
Alkyne	3300
Arene	3030
Aldehyde	2900–2820 and 2775–2700
C—H bending variations	
Alkane	1485–1365
Arene 5 adjacent hydrogen atoms	750 and 700
4 adjacent hydrogen atoms	750
3 adjacent hydrogen atoms	780
2 adjacent hydrogen atoms	830
1 adjacent hydrogen atom	880
N—H stretching vibrations	
Amine	3500–3300
Amide	3500–3140
O—H stretching vibrations	
Alcohols and phenols	3750–3200
Carboxylic acids	3300–2500
C=C stretching vibrations	
Isolated alkene	1669–1645
Arene	1600, 1580, 1500, 1450
C=O stretching vibrations	
Aldehydes, saturated alkyl	1740–1720
Ketones alkyl	1720–1700
Ketones aryl	1700–1680
Carboxylic acids alkyl	1725–1700
aryl	1700–1680
Carboxylic acid anhydrides	1850–1800 and 1790–1740
Acyl halides chlorides	1795
bromides	1810
Esters, saturated	1750–1735
Amides	1700–1630
Triple bond stretching vibrations	
CN	2260–2215
CC	2260–2100

^{1}H nuclear magnetic resonance chemical shifts relative to tetramethylsilane (TMS)

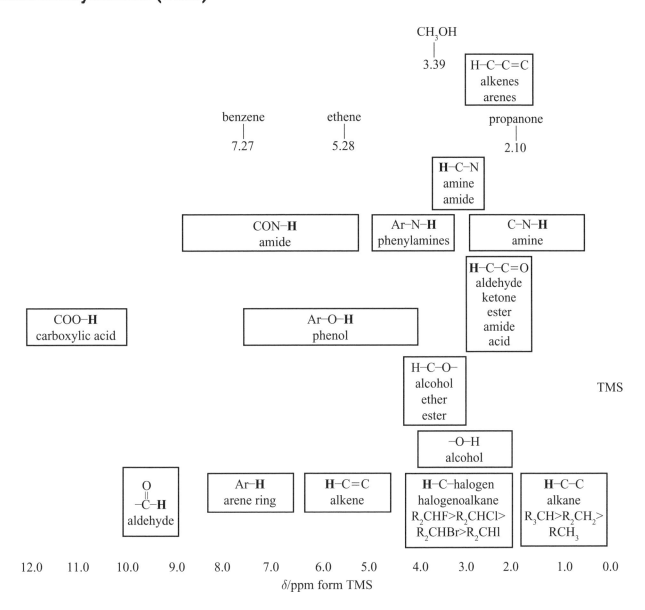

^{13}C nuclear magnetic resonance chemical shifts relative to tetramethylsilane (TMS)

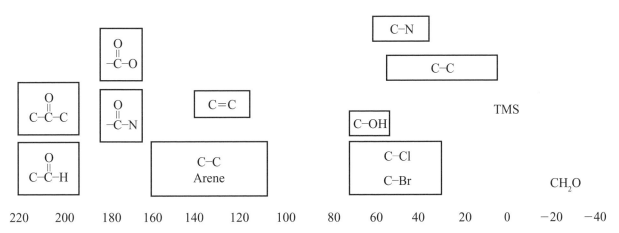

Pauling electronegativity index

												H						He
												2·1						
Li	Be											B	C	N	O	F	Ne	
1·0	1·5											2·0	2·5	3·0	3·5	4·0		
Na	Mg											Al	Si	P	S	Cl	Ar	
0·9	1·2											1·5	1·9	2·1	2·5	3·0		
K	Ca	Sc	Ti	V	Cr	Mn	Fe	Co	Ni	Cu	Zn	Ga	Ge	As	Se	Br	Kr	
0·8	1·0	1·3	1·5	1·6	1·6	1·5	1·8	1·8	1·8	1·9	1·6	1·6	2·0	2·0	2·4	2·8		
Rb	Sr	Y	Zr	Nb	Mo	Tc	Ru	Rh	Pd	Ag	Cd	In	Sn	Sb	Te	I	Xe	
0·8	1·0	1·2	1·3	1·6	2·1	1·9	2·2	2·2	2·2	1·9	1·6	1·7	1·9	1·9	2·1	2·5		
Cs	Ba	La	Hf	Ta	W	Re	Os	Ir	Pt	Au	Hg	Tl	Pb	Bi	Po	At	Rn	
0·7	0·9	1·1	1·3	1·5	2·3	1·9	2·2	2·2	2·2	2·5	2·0	1·6	1·8	1·9	2·0	2·2		

Relation in electronegativity difference, ΔN_e and ionic character P/%

Electronegativity difference ΔN_e	0·1	0·3	0·5	0·7	1·0	1·3	1·5	1·7	2·0	2·5	3·0
Percentage ionic character $P/\%$	0·5	2	6	12	22	34	43	51	63	79	89

Indicators

		pK_{in} (at 298 K)	*acid*	pH range	*alkaline*
1	Thymol blue (acid)	1.7	red	1.2–2.8	yellow
2	Screened methyl orange	3.7	purple	3.2–4.2	green
3	Methyl orange	3.7	red	3.2–4.4	yellow
4	Bromophenol blue	4.0	yellow	2.8–4.6	blue
5	Bromocresol green	4.7	yellow	3.8–5.4	blue
6	Methyl red	5.1	red	4.2–6.3	yellow
7	Litmus		red	5.0–8.0	blue
8	Bromothymol blue	7.0	yellow	6.0–7.6	blue
9	Phenol red	7.9	yellow	6.8–8.4	red
10	Phenolphthalein (in ethanol)	9.3	colourless	8.2–10.0	red

Standard electrode potentials

$E^{\ominus}$ Standard electrode potential of aqueous system at 298 K, that is, standard emf of electrochemical cell in the hydrogen half-cell forms the left-hand side electrode system.

	Right-hand electrode system	$E^{\ominus}/V$
1	$Na^+ + e^- \rightleftharpoons Na$	-2.71
2	$Mg^{2+} + 2e^- \rightleftharpoons Mg$	-2.37
3	$Al^{3+} + 3e^- \rightleftharpoons Al$	-1.66
4	$V^{2+} + 2e^- \rightleftharpoons V$	-1.18
5	$Zn^{2+} + 2e^- \rightleftharpoons Zn$	-0.76
6	$Cr^{3+} + 3e^- \rightleftharpoons Cr$	-0.74
7	$Fe^{2+} + 2e^- \rightleftharpoons Fe$	-0.44
8	$Cr^{3+} + e^- \rightleftharpoons Cr^{2+}$	-0.41
9	$V^{3+} + e^- \rightleftharpoons V^{2+}$	-0.26
10	$Ni^{2+} + 2e^- \rightleftharpoons Ni$	-0.25
11	$H+ + e^- \rightleftharpoons \frac{1}{2}H_2$	0.00
12	$\frac{1}{2}S_4O_6^{2-} + e^- \rightleftharpoons S_2O_3^{2-}$	$+0.09$
13	$Cu^{2+} + e^- \rightleftharpoons Cu^+$	$+0.15$
14	$Cu^{2+} + 2e^- \rightleftharpoons Cu$	$+0.34$
15	$VO^{2+} + 2H^+ + e^- \rightleftharpoons V^{3+} + H_2O$	$+0.34$
16	$\frac{1}{2}O_2 + H_2O + 2e^- \rightleftharpoons 2OH^-$	$+0.40$
17	$S_2O_3^{2-} + 6H^+ + 4e^- \rightleftharpoons 2S + 3H_2O$	$+0.47$
18	$Cu^+ + e^- \rightleftharpoons Cu$	$+0.52$
19	$\frac{1}{2}I_2 + e^- \rightleftharpoons I^-$	$+0.54$
20	$3O_2 + 2H^+ + 2e^- \rightleftharpoons H_2O_2$	$+0.68$
21	$Fe^{3+} + e^- \rightleftharpoons Fe^{2+}$	$+0.77$
22	$Ag^+ + e^- \rightleftharpoons Ag$	$+0.80$
23	$NO_3^- + 2H^+ + e^- \rightleftharpoons NO_2 + H_2O$	$+0.80$
24	$ClO^- + H_2O + 2e^- \rightleftharpoons Cl^- + 2OH^-$	$+0.89$
25	$VO2^+ + 2H^+ + e^- \rightleftharpoons VO^{2+} + H_2O$	$+1.00$
26	$\frac{1}{2}Br_2 + e^- \rightleftharpoons Br^-$	$+1.09$
27	$\frac{1}{2}O_2 + 2H^+ + 7H^+ + 3e^- \rightleftharpoons Cr^{3+} + \frac{7}{2}H_2O$	$+1.23$
28	$\frac{1}{2}Cr_2O_7^{2-} + 2e^- \rightleftharpoons H_2O$	$+1.33$
29	$\frac{1}{2}Cl_2 + e^- \rightleftharpoons Cl^-$	$+1.36$
30	$MnO4^- + 8H^+ + 5e^- \rightleftharpoons Mn^{2+} + 4H_2O$	$+1.51$
31	$\frac{1}{2}H_2O_2 + H_+ + e^- \rightleftharpoons H_2O$	$+1.77$

Answers

1. Atomic structure and isotopes

1 $^{20}_{10}$Ne *1 mark for 20, 1 mark for Ne and 10*

2 15 protons **(1)**, 16 neutrons **(1)**, 18 electrons **(1)**

3 (a) They have the same atomic number/number of protons **(1)** but different mass numbers/numbers of neutrons. **(1)**

    (b) neutron **(1)**

4 (a) The number of protons in the nucleus. **(1)**

       They may have different numbers of neutrons. **(1)**

    (b) Uncharged atoms have equal numbers of protons and electrons. **(1)**

       The relative charge on a proton is +1 and the relative charge on an electron is –1 (and neutrons are neutral). **(1)**

2. Mass spectrometry

1 The weighted mean mass of an atom of an element **(1)** compared to $\frac{1}{12}$th the mass of a ^{12}C atom. **(1)**

2 Peak at *m/z* 70 due to $(^{35}Cl–^{35}Cl)^+$ **(1)**

   Peak at *m/z* 72 due to $(^{35}Cl–^{37}Cl)^+$ **(1)**

   Peak at *m/z* 74 due to $(^{37}Cl–^{37}Cl)^+$ **(1)**

3. Shells, sub-shells and orbitals

1 (a) A region around the nucleus where there is a high probability of finding an electron **(1)** and which can hold up to two electrons with opposite spins. **(1)**

    (b) *1 mark for each correct diagram*:

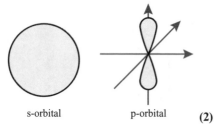

       s-orbital   p-orbital   **(2)**

2 (a) 2 in s, 6 in p, 10 in d **(1)**

    (b) 2 in shell 1, 8 in shell 2, 18 in shell 3, 32 in shell 4 **(1)**

3 The box represents the orbital. **(1)**

   Each arrow represents an electron with its spin. **(1)**

4. Electronic configurations

1 (a) $1s^2\,2s^2\,2p^6\,3s^2\,3p^2$ **(1)**

    (b) $1s^2\,2s^2\,2p^6\,3s^2\,3p^6\,3d^6\,4s^2$ **(1)**

    (c) $1s^2\,2s^2\,2p^6\,3s^2\,3p^6\,3d^{10}\,4s^2\,4p^5$ **(1)**

2 (a) Period 2, Group 6, p-block (all correct) **(1)**

    (b) 1s 2s 2p$_x$ 2p$_y$ 2p$_z$ **(1)**

       ↑↓ ↑↓ ↑↓ ↑ ☐

5. Ionisation energies

(a) Group 5 because the largest increase in first ionisation energy is between the 5th and 6th ionisation energies. **(1)**

(b) 4th, 5th, 6th, 7th

   all four correct = 2 marks / three correct = 1 mark

   If more than four answers are given, deduct one mark for each extra answer to 0 marks minimum.

6. Periodicity

1 *1 mark for each correct row to 3 maximum*:

Element	Structure	Bonding
lithium	giant lattice	metallic
boron	giant lattice	covalent
oxygen	simple molecular	covalent

2 **C (1)**

3 krypton **(1)**

4 In selenium, an electron is removed from a 4p-orbital containing two electrons. **(1)**

   These electrons repel each other more strongly than the ones in arsenic, so the electron is lost more easily. **(1)**

8. Ions

1 **A (1)**

   Nickel(II) ions, Ni^{2+}(aq), are green but sulfate ions are colourless.

2 (a) 2.8.8 **(1)**

    (b) 2.8.8 **(1)**

    (c) 2.8.8 **(1)**

    (d) 2.8 **(1)**

3

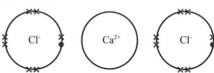

   correct number of outer electrons **(1)**

   correct formulae and charges of the ions **(1)**

   1:2 ratio of Ca^{2+} and Cl$^-$ ions **(1)**

9. Ionic bonds

1 (a) S^{2-} **(1)**

    (b) The ions are isoelectronic. **(1)**

       S^{2-} has the lowest atomic number/smallest nuclear charge. **(1)**

       S^{2-} has the weakest attraction between its nucleus and its electrons. **(1)**

2 (a) The ionic radius increases going down the group. **(1)**

    (b) There are more occupied shells as you go down a group. **(1)**

3 The sizes of oxide ions and fluoride ions is similar. **(1)**

   Oxide ions have a higher charge / charge density. **(1)**

   Oxide ions are more strongly attracted to magnesium ions. **(1)**

   Answers could refer to fluoride ions and their lower charge density instead.

10. Covalent bonds

1 (Electrostatic attraction between two nuclei and the) shared pair **(1)** of electrons (between them). **(1)**

2 (a) H   **(1)**

       H ⦂ C ⦂ H

          H

    (b) H ⦂ O ⦂ H **(1)**

    (c) :Cl: **(1)**

       :Cl⦂ B ⦂Cl:

    Diagrams may be shown with or without circles.

3 (a) dative covalent bond **(1)**

(b) O atom donates a lone pair of electrons to the bond. **(1)**
H⁺ ion has a vacant orbital. **(1)**

11. Covalent bond strength

1 (a) Bond length should decrease N–N > N=N > N≡N **(1)**
Bond strength should increase N–N < N=N < N≡N **(1)**

(b) The electrostatic attraction between the two nuclei and the shared electrons increases as more pairs of electrons are shared. **(1)**

2 The nuclear charge in an O atom is greater than the nuclear charge in an N atom. **(1)**

3 Ö⦂C⦂Ö

correct number of shared electrons **(1)** *correct number of lone pairs of electrons* **(1)**
Diagram may be shown with or without circles.

12. Shapes of molecules and ions

1 (a) 109.5° **(1)**
4 bonding pairs of electrons around the central atom

(b) 109.5° **(1)**
4 bonding pairs of electrons around the central atom

(c) 104.5° **(1)**
2 bonding pairs and 2 lone pairs around the central atom

(d) 180° **(1)**
2 bonding pairs around the central atom

(e) 90° *and* 120° **(1)**
5 bonding pairs around the central atom

2 BH_3 120° **(1)**
PH_3 107° **(1)**
BH_3 has 3 bonding pairs and no lone pairs but PH_3 has three bonding pairs and one lone pair, and lone pairs repel more than bonding pairs. **(1)**

3 trigonal planar **(1)** 120° **(1)**

13. Electronegativity and bond polarity

1 (a) The ability of an atom to attract the bonding electrons in a covalent bond. **(1)**

(b) Increase across a period. **(1)**
Decrease down a group. **(1)**

2 The electronegativity of the halogens increases as you go up the group. **(1)**
The larger the difference in electronegativity between hydrogen and halogen, the greater the ionic character. **(1)**

3 CBr_4 **(1)**

14. London forces

1 Instantaneous dipoles arise in molecules (because of fluctuations in the electron cloud). **(1)**
These dipoles induce dipoles in neighbouring molecules. **(1)**
There are attractive forces between the partial positive and partial negative charges. **(1)**

2 The number of electrons / M_r increases going from chlorine through bromine to iodine. **(1)**
The London forces increase $Cl_2 < Br_2 < I_2$. **(1)**
Chlorine is a gas because it has the weakest London forces / iodine is a solid because it has the strongest London forces. **(1)**

3 D, because it has the greatest number of points of contact / surface area. **(1)**

15. Permanent dipoles and hydrogen bonds

1 Ammonia has the higher boiling temperature. **(1)**
Ammonia can form hydrogen bonds but methane cannot. Additional energy is needed to overcome the intermolecular forces in ammonia. **(1)**

2 (a) Fluorine is more electronegative than bromine. **(1)**
HF is more polar than HBr. **(1)**

(b)
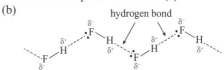

Hydrogen bond shown between H on one molecule and lone pair of electrons on F of another molecule. **(1)**
FHF atoms arranged in a straight line. **(1)**

16. Choosing a solvent

1 D **(1)**

2 (a) London forces form between the alkane molecules and the CH_3 groups in propanone. **(1)**

(b) Hydrogen bonds form between water molecules and oxygen atom in propanone. **(1)**

17. Giant lattices

1 (a) giant metallic lattice **(1)**
strong electrostatic force of attraction between positive ions **(1)** and delocalised electrons **(1)**

(b) giant ionic lattice **(1)**
strong electrostatic force of attraction between (positive) sodium ions **(1)** and (negative) chloride ions **(1)**

(c) simple molecular lattice **(1)**
covalent bonds between iodine atoms **(1)**
London forces between molecules **(1)**

2 giant covalent lattice **(1)**
covalent bonds **(1)** between silicon atoms and oxygen atoms **(1)**

18. Structure and properties

A A giant metallic lattice structure with metallic bonding, **(1)**
it conducts electricity when solid or molten. **(1)**

B Giant ionic lattice structure with ionic bonding, **(1)**
it conducts electricity when molten but not when solid. **(1)**

C Simple molecular structure with weak London forces between molecules, **(1)**
it has low melting and boiling temperatures (and does not conduct when solid or molten). **(1)**

D Giant covalent lattice structure with covalent bonding, **(1)**
it has very high melting and boiling temperatures *and* does not conduct when solid or molten. **(1)**

20. Oxidation numbers

1 (a) 0 **(1)**
(b) –1 **(1)**
(c) –1 **(1)**
(d) +4 **(1)**
(e) +7 **(1)**

2 (a) –2 **(1)**
(b) +3 **(1)**
(c) +3 **(1)**
(d) –3 **(1)**
(e) +3 **(1)**

3 (a) phosphorus(V) chloride **(1)**
(b) sodium chromate(VI) **(1)**
(c) vanadate(III) ion **(1)**

4 (a) $FeCl_3$ **(1)**

(b) PbO_2 **(1)**

(c) NH_4VO_3 **(1)**

21. Redox reactions

The oxidation number of carbon increases from +2 in CO to +4 in CO_2 **(1)** so this is oxidation. **(1)**
The oxidation number of nitrogen decreases from +2 in NO to 0 in N_2 **(1)** so this is reduction. **(1)**

22. Ionic half-equations

1 (a) $Ag^+ + Cu \rightarrow 2Ag + Cu^{2+}$ **(1)**

(b) $MnO_4^- + 8H^+ + 5Fe^{2+} \rightarrow Mn^{2+} + 4H_2O + 5Fe^{3+}$ **(1)**

2 $2H^+ + 2e^- \rightarrow H_2$ **(1)**

3 $H_2O_2 \rightarrow O_2 + 2H^+ + 2e^-$ **(1)**

23. Reactions of Group 2 elements

1 (a) *For each equation, 1 mark for correct species and balancing, 1 mark for correct state symbols:*
$2Mg(s) + O_2(g) \rightarrow 2MgO(s)$
$Mg(s) + H_2O(g) \rightarrow MgO(s) + H_2(g)$

(b) There is very slow bubbling (due to hydrogen gas). **(1)**
$Mg(s) + 2H_2O(l) \rightarrow Mg(OH)_2(s) + H_2(g)$
1 mark for correct species and balancing, 1 mark for correct state symbols

2 Reactivity increases. **(1)**
Group 2 metals form M^{2+} ions in reactions. **(1)**
Ionisation energies decrease (down the group). **(1)**

3 Barium is stored under oil. **(1)**
To prevent it reacting with air / oxygen / water (because it is very reactive). **(1)**

24. Reactions of Group 2 compounds

(a) $CaO(s) + H_2O(l) \rightarrow Ca(OH)_2(aq)$
1 mark for correct species and balancing, 1 mark for correct state symbols

(b) Mixture turns blue **(1)** because limewater is alkaline / pH>7 **(1)**

(c) $Ca(OH)_2(aq) + 2HCl(aq) \rightarrow CaCl_2(aq) + 2H_2O(l)$
1 mark for correct species and balancing, 1 mark for correct state symbols

25. Stability of carbonates and nitrates

(a) Lithium nitrate decomposes at a lower temperature / potassium nitrate decomposes at a higher temperature. **(1)**
The products are different (lithium oxide, nitrogen dioxide and oxygen / potassium nitrite and oxygen). **(1)**
$4LiNO_3(s) \rightarrow 2Li_2O(s) + 4NO_2(g) + O_2(g)$ **(1)**
$2KNO_3(s) \rightarrow 2KNO_2(s) + O_2(g)$ **(1)**
State symbols need not be shown as they were not asked for, but should be correct if given

(b) Lithium ions are smaller than potassium ions / have a higher charge density. **(1)**
Lithium ions have a greater (polarising) effect on the nitrate ion / N–O bond. **(1)**
The reverse argument for potassium is acceptable.

26. Flame tests

1 Electrons are excited (by heat energy). **(1)**
Electrons are promoted to higher energy levels. **(1)**
Light is emitted as the electrons return to a lower energy state / the ground state. **(1)**

2 C **(1)**

27. Properties of Group 7 elements

1 Answer could include:
(dark) solid **(1)**
electronegativity less than that of iodine **(1)**
because melting and boiling temperatures increase and electronegativity values decrease down the group **(1)**

2 simple molecular / molecular crystal **(1)**
covalent bonds between atoms **(1)**
london forces between molecules **(1)**
weak London forces / easily overcome / little energy needed to break them **(1)**

28. Reactions of Group 7 elements

(a) The solution turns yellow / orange / pale brown (due to bromine). **(1)**
Two layers are produced and the top layer is orange. **(1)**

(b) $Cl_2 + 2Br^- \rightarrow 2Cl^- + Br_2$ **(1)**

(c) Oxidising agent **(1)** because it gains electrons (from bromine). **(1)**

29. Reactions of chlorine

(a) sodium iodate(V) **(1)**

(b) disproportionation **(1)**
Because iodine is simultaneously reduced to iodide ions $(0 \rightarrow -1)$ **(1)** and oxidised to iodate(V) ions. $(0 \rightarrow +5)$ **(1)**

30. Halides as reducing agents

1 $H_2SO_4 + 8H^+ + 8e^- \rightarrow 4H_2O + H_2S$ **(1)**

2 $8HI + H_2SO_4 \rightarrow 4H_2O + H_2S + 4I_2$
1 mark for correct species, 1 mark for correct balancing.

31. Other reactions of halides

1 (a) Silver fluoride is soluble / does not form a precipitate. **(1)**

(b) (i) Silver chloride forms a white precipitate. **(1)**
(ii) Add dilute ammonia. **(1)**
If the precipitate dissolves it shows chloride ions present. **(1)**

2 The reaction produces hydrogen bromide **(1)** which forms ammonium bromide with ammonia. **(1)**
$NH_3(g) + HBr(g) \rightarrow NH_4Br(s)$ **(1)**

33. Moles and molar mass

1 (a) $2 \times (2 + 6) = 2 \times 8 = 16$ mol **(1)**

(b) $16 \times 6.02 \times 10^{23} = 9.632 \times 10^{24}$ **(1)**

2 Idea that the mole is the unit for amount of substance but the Avogadro constant is the number of particles in one mole of particles. **(1)**
The word 'entity' is sometimes used to represent particles in the definition of the mole.

3 (a) $0.5 \times 18.0 = 9.0$ g **(1)**

(b) M_r of $C_2H_5OH = (2 \times 12.0) + (6 \times 1.0) + (1 \times 16.0)$
$= 46.0$ **(1)**
molar mass = 46.0 g mol^{-1}
$1.25 \times 46.0 = 57.5$ g **(1)**

4 4.5 g of water because:
4.5 g of water contains: $3 \times 4.5 \div 18.0$
$= 0.75$ mol of atoms **(1)**
3.8 g of ethanol contains: $9 \times 3.8 \div 46.0$
$= 0.743$ mol of atoms **(1)**

34. Empirical and molecular formulae

1 Molar mass of $NO_2 = 14.0 + (2 \times 16.0) = 46.0$
Factor $= 92.0 \div 46.0 = 2$
Molecular formula is N_2O_4 **(1)**

2 Oxygen $= 100 - (32.4 + 22.6) = 45.0\%$ **(1)**

Na	S	O
$32.4 \div 23.0 = 1.41$	$22.6 \div 32.1 = 0.704$	$45.0 \div 16.0 = 2.81$ **(1)**
$1.41 \div 0.704 = 2$	$0.704 \div 0.704 = 1$	$2.81 \div 0.704 = 4$

Empirical formula is Na_2SO_4 **(1)**

35. Reacting masses calculations

1 $Mg + 2HCl \rightarrow MgCl_2 + H_2$
molar mass of $Mg = 24.3$ g mol^{-1} and molar mass of $MgCl_2 = 95.3$ g mol^{-1} **(1)**
amount of magnesium $= 5.98 \div 24.3 = 0.246$ mol **(1)**
amount of magnesium chloride $= 0.246$ mol **(1)**
mass of magnesium chloride $= 246 \times 95.3 = 23.4$ g **(1)**

2 molar mass of $Cu = 63.5$ g mol^{-1} and molar mass of $H_2O = 18.0$ g mol^{-1} **(1)**
amount of copper $= 3.51 \div 63.5 = 0.0553$ mol <u>and</u>
amount of water $= 0.499 \div 18.0 = 0.0277$ mol **(1)**
ratio is $0.0553 : 0.0277 = 2 : 1$ **(1)**
equation must be $Cu_2O + H_2 \rightarrow 2Cu + H_2O$ **(1)**

36. Gas volume calculations

(a) amount of carbon $= 0.536 \div 12.0 = 0.0447$ mol **(1)**
volume of oxygen $= 0.0447 \times 22.4 = 1.00$ dm^3 **(1)**
(b) 1.40 dm^3 **(1)**
1.00 dm^3 is CO_2 and 0.40 dm^3 is excess O_2.

37. Concentrations of solutions

1 (a) volume $= 50 \div 1000 = 0.05$ dm^3 **(1)**
concentration $= 0.1 \div 0.05 = 2$ mol dm^{-3} **(1)**
(b) $2 \times 0.1 = 0.2$ mol **(1)**

2 C **(1)**
$250 \div 1000 \times 0.1 \times 180.0 = 4.5$ g

38. Doing a titration

1 yellow to orange **(1)**

2 24.85 <u>and</u> 24.95 ticked **(1)**

39. Titration calculations

Amount of $NH_2SO_3H = 0.080 \times 23.45 \times 10^{-3}$
$\qquad\qquad\qquad = 1.876 \times 10^{-3}$ mol **(1)**
Amount of $NaOH = 1 \times 1.876 \times 10^{-3} = 1.876 \times 10^{-3}$ mol **(1)**
Concentration of $NaOH = (1.876 \times 10^{-3}) \div (25.00 \times 10^{-3})$
$\qquad\qquad\qquad = 0.075$ mol dm^{-3} **(1)**

40. Atom economy

(a) molar mass of $NaNO_3 = 85.0$ g mol^{-1}
molar mass of $AgCl = 143.4$ g mol^{-1}
sum of molar masses $= 85.0 + 143.4 = 228.4$ g mol^{-1} **(1)**
atom economy $= 100 \times 143.4 \div 228.4 = 62.8\%$ **(1)**
(b) molar mass of $AgNO_3 = 169.9$ g mol^{-1}
theoretical yield $= 20.0 \div 169.9 \times 143.4 = 16.88$ g **(1)**
percentage yield $= 12.0 \div 16.88 \times 100 = 71.1\%$ **(1)**

42. Alkanes

(a) molecular formula: C_8H_{18} **(1)**
empirical formula: C_4H_9 **(1)**
structural formula: $CH_3(CH_2)_6CH_3$ or
$CH_3CH_2CH_2CH_2CH_2CH_2CH_2CH_3$ **(1)**
(b) displayed formula **(1)**

No 'empty sticks', i.e. all hydrogen atoms must be shown.
skeletal formula **(1)**

43. Isomers of alkanes

1 mark for each correct structure, 1 mark for each correct name. Branches may be shown above or below the main chain, and may be drawn from the right (rather than from the left as shown here).

1 c–c–c–c–c–c–c

2 c–c–c–c–c–c
     |
     c

3 c–c–c–c–c–c
       |
       c

4   c
     |
  c–c–c–c–c
     |
     c

5   c
      |
  c–c–c–c–c
      |
      c

6 c–c–c–c–c
    | |
    c c

7 c–c–c–c–c
   | |
   c   c

8 c–c–c–c–c
     |
     c
     |
     c

9   c
     |
  c–c–c–c
    | |
    c c

1 heptane
2 2-methylhexane
3 3-methylhexane
4 2,2-dimethylpentane
5 3,3-dimethylpentane
6 2,3-dimethylpentane
7 2,4-dimethylpentane
8 3-ethylpentane
9 2,3,3-trimethylbutane

44. Alkenes

1 molecular formula: C_8H_{16} **(1)**
empirical formula: CH_2 **(1)**

2 displayed formula **(1)**

No 'empty sticks', i.e. all hydrogen atoms must be shown.
skeletal formula **(1)**

45. Isomers of alkenes

1 3-methylbut-1-ene **(1)**

2 *1 mark for each structure, with 1 mark for each correct name*
E-pent-2-ene

Z-pent-2-ene

46. Using crude oil

Fractional distillation involves heating crude oil, then cooling and condensing the vapours at different temperatures. **(1)**

It is carried out to separate the crude oil into fractions (which each contain hydrocarbons with similar boiling points). **(1)**

Cracking involves passing the vapours from fractions containing larger alkane molecules over a hot (zeolite) catalyst, which produces smaller alkane molecules and alkenes. **(1)**

It is carried out to match the supply of each fraction with its demand (and to produce alkenes to make polymers). **(1)**

Reforming involves heating fractions in the presence of catalysts to produce branched hydrocarbons and cyclic hydrocarbons. **(1)**

These hydrocarbons burn more efficiently in engines / have a higher octane rating . **(1)**

47. Hydrocarbons as fuels

1   $C_2H_6 + 2O_2 \rightarrow CO + 3H_2O + C$ **(1)**

2   Carbon monoxide is toxic **(1)**
Sulfur dioxide and nitrogen oxides are acidic **(1)** and cause acid rain. **(1)**

48. Alternative fuels

1   (a) The amount of carbon dioxide released in its manufacture and use **(1)** is equal to the amount of carbon dioxide absorbed for photosynthesis. **(1)**

(b) *Sensible reason why a fossil fuel may have been involved in the manufacture or use of bioethanol for **1** mark, e.g. fossil fuel used to:*

  • make the fertilisers to grow the crops
  • run the equipment to make the bioethanol, e.g. tractors, distillation column
  • transport the bioethanol, e.g. tankers.

2   *One environmental reason for using alternative fuels for **1** mark, e.g.*

  • less carbon dioxide released / smaller (overall) emissions of greenhouse gases
  • conservation of finite / non-renewable resources.

49. Halogenoalkanes and alcohols

1   *The hydroxyl and methyl groups may be shown above or below (the hydroxyl group may be shown left or right if it is on the end).*
1 mark for each correct structure.

(a)

(b)

(c)

50. Substitution reactions of alkanes

1   A radical is a species with an unpaired electron. **(1)**
It is formed by homolytic fission of a covalent bond. **(1)**

2   Initiation: $Br_2 \rightarrow \cdot Br + \cdot Br$ **(1)**
Propagation: $CH_3CH_3 + \cdot Br \rightarrow \cdot CH_2CH_3 + HBr$ **(1)**
$\quad\quad\quad\quad \cdot CH_2CH_3 + Br_2 \rightarrow CH_3CH_2Br + \cdot Br$ **(1)**
Termination: $\cdot CH_2CH_3 + \cdot Br \rightarrow CH_3CH_2Br$
$\quad\quad\quad\quad \cdot CH_2CH_3 + \cdot CH_2CH_3 \rightarrow C_4H_{10}$
$\quad\quad\quad\quad \cdot Br + \cdot Br \rightarrow Br_2$

*One for **1** mark*

2

The halogen atoms and methyl groups may be shown above or below (the halogen atom may be shown left or right if it is on the end).
1 mark for each correct structure.

(a)   (b)

(c)

51. Alkenes and hydrogen halides

(a) electrophilic addition **(1)**
Both words needed.

(b)

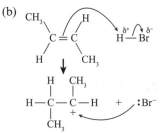

1 mark for each curly arrow
Make sure they begin on a bond or next to the lone pair of electrons on the bromide ion.
1 mark for correct structure of the carbocation
The methyl groups can be attached to any of the bonds on each carbon atom.

52. More addition reactions of alkenes

(a) Major product: 2-bromopropane **(1)**
Minor product: 1-bromopropane **(1)**

(b)

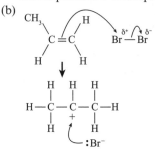

1 mark for each curly arrow
Make sure they begin on a bond or next to the lone pair of electrons on the bromide ion.
1 mark for correct structure of the (secondary) carbocation.

54. Alkenes and alcohols

(a) $CH_3CH=CHCH_3 + H_2O \rightarrow CH_3CH_2CH(OH)CH_3$ **(1)**
   butan-2-ol **(1)**
(b) $CH_3CH=CHCH_3 + [O] + H_2O$
   $\rightarrow CH_3CH(OH)CH(OH)CH_3$ **(1)**
   butane-2,3-diol **(1)**

55. Addition polymerisation

(a)

(b) chloroethene/1-chloroethene **(1)**

(c)
```
  ┌ H   H ┐
  │ |   | │
──┼ C ─ C ┼──
  │ |   | │
  └ H   Cl┘   (1)
```

(d) addition polymer **(1)**

56. Polymer waste

1 for recycling **(1)**
   for incineration to release energy **(1)**
   for use as a chemical feedstock for cracking. **(1)**
2 Cracking polymers **(1)** to produce a fraction for use in making new polymers. **(1)**
3 hydrogen chloride / nitrogen oxides / sulfur dioxide / carbon monoxide **(1)**
   Carbon dioxide is produced but is not toxic.
   Toxic 'dioxins' are produced but these are solids at room temperature.

57. Alcohols from halogenoalkanes

1 A species that can donate a lone pair of electrons to form a dative covalent bond. **(1)**
2 *1 mark for each correct curly arrow*

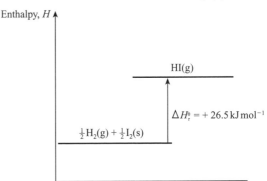

The minus sign on the OH⁻ ion should be shown over the O atom in the mechanism.
Organic product is: ethanol. **(1)**

58. Reactivity of halogenoalkanes

D **(1)**
The mean bond enthalpy for the C–F bond is $467\,kJ\,mol^{-1}$.

59. More halogenoalkane reactions

Organic product: ethylamine / aminoethane / ethanamine **(1)**
Reaction mechanism **(4)**:

60. Oxidation of alcohols

(a) Reflux:
   $CH_3CH_2CH_2CH_2OH + 2[O] \rightarrow CH_3CH_2CH_2COOH + H_2O$
   1 mark for correct organic product
   1 mark for correctly balanced
   Distillation:
   $CH_3CH_2CH_2CH_2OH + [O] \rightarrow CH_3CH_2CH_2CHO + H_2O$
   1 mark for correct species and balanced
(b) $CH_3CH_2CH(OH)CH_3 + [O] \rightarrow CH_3CH_2COCH_3 + H_2O$
   1 mark for correct species and balanced

61. Halogenoalkanes from alcohols

Chlorination: phosphorus(V) chloride / PCl_5 **(1)**
Bromination: potassium bromide <u>and</u> 50% concentrated sulfuric acid **(1)**
Iodination: red phosphorus <u>and</u> iodine **(1)**

63. Structures from mass spectra

(a) $m/z = 15$ – due to CH_3^+ **(1)**
   $m/z = 29$ – due to $CH_3CH_2^+$ **(1)**
   $m/z = 31$ – due to CH_2OH^+ **(1)**
(b) $(CH_3CH_2OH)^+ \rightarrow CH_3^+ + \cdot CH_2OH$ **(1)**
   $(CH_3CH_2OH)^+ \rightarrow CH_3CH_2^+ + \cdot OH$ **(1)**
   $(CH_3CH_2OH)^+ \rightarrow CH_2OH^+ + \cdot CH_3$ **(1)**

64. Infrared spectroscopy

(a) A could be due to the O–H group in an alcohol. **(1)**
   The peak is at about $3350\,cm^{-1}$
   B could be due to the C–H group in an alkane. **(1)**
   The peak is at about $2950\,cm^{-1}$
(b) The compound could be propan-1-ol / propan-2-ol. **(1)**
   Its formula would be $CH_3OCH_2CH_2OH$ / $CH_3CH(OH)CH_3$. **(1)**
   This contains C–H bonds and an O–H bond, and matches the molecular formula given. **(1)**

65. Enthalpy changes

$\frac{1}{2}H_2(g) + \frac{1}{2}I_2(s) \rightarrow HI(g)$ **(1)**
The equation must not be doubled to remove the fractions and it must have correct state symbols.
2 marks for the enthalpy level diagram:
• correct position and labelling of reactants and products **(1)**
• correct direction of arrow and labelling **(1)**

```
Enthalpy, H ▲
                                    HI(g)
                                  ──────────
                                      │
                                      │  ΔH°_r = + 26.5 kJ mol⁻¹
                                      │
  ½H₂(g) + ½I₂(s)                     │
  ──────────────                      │
```

66. Measuring enthalpy changes

$Q = 40.0 \times 4.18 \times 35.0 = 5852\,J$ **(1)**
Energy change for copper(II) sulfate $= -5.852\,kJ$
Amount of copper(II) sulfate $= 1.00 \times 0.040 = 0.040\,mol$ **(1)**
$\Delta H_r = (-5.852) / 0.040 = -146\,kJ\,mol^{-1}$
1 mark for 146, 1 mark for – sign

67. Enthalpy cycles

(a) *1 mark for correct cycle*

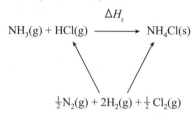

$$NH_3(g) + HCl(g) \xrightarrow{\Delta H_r} NH_4Cl(s)$$

$$\tfrac{1}{2}N_2(g) + 2H_2(g) + \tfrac{1}{2}Cl_2(g)$$

(b) *1 mark for correct cycle*

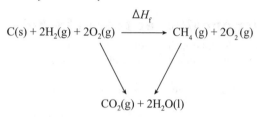

$$C(s) + 2H_2(g) + 2O_2(g) \xrightarrow{\Delta H_f} CH_4(g) + 2O_2(g)$$

$$CO_2(g) + 2H_2O(l)$$

68. Using enthalpy cycles

$\Delta H = (\sum \Delta H_f \text{[products]}) - (\sum \Delta H_f \text{[reactants]})$
= $2(-393.5) + 3(-285.8) - (-277.1)$ **(1)**
= $-787.0 - 857.4 + 277.1$
= $-1367.3 \, \text{kJ mol}^{-1}$ **(1)**
Answer must have correct sign and unit.
Standard enthalpy of formation of oxygen is zero.

69. Mean bond enthalpy calculations

1   H_2O is present as a liquid in the equation but bond enthalpies are for bonds broken in the gaseous state. **(1)**
The data book value is the mean from many different molecules (not just for CH_4). **(1)**

2   $\sum(\text{bonds broken}) = (2 \times 436) + 498 = 1370 \, \text{kJ mol}^{-1}$ **(1)**
$\sum(\text{bonds made}) = 4 \times 463 = 1852 \, \text{kJ mol}^{-1}$ **(1)**
Enthalpy change = $1370 - 1852 = -482 \, \text{kJ mol}^{-1}$ **(1)**

70. Changing reaction rate

Decrease in concentration:
- fewer solute particles (in the same volume) **(1)**
- fewer collisions / reduced frequency of collision between reactant particles **(1)**

Change in size:
- surface area decreased **(1)**
- fewer collisions / reduced frequency of collision between reactant particles **(1)**

71. Maxwell–Boltzmann model

1 mark for each of the following features:
- axes labelled as shown
- one curve with sensible asymmetrical shape as shown
- second curve drawn appropriately as shown (and labelled)
- activation energy indicated
- shaded areas to the right of the E_a line.

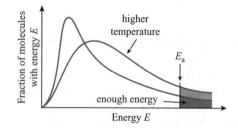

72. Catalysts

1   *1 mark for each of the following features:*
- Maxwell–Boltzmann curve, starts at origin and does not cross axis
- sensible shape
- axes labelled
- activation energies with and without catalyst shown
- shaded areas to represent the numbers of molecules with $\geqslant {}^3E_a$

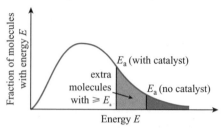

2   *1 mark for each of the following features:*
- reaction profile with enthalpy of products higher than reactants
- curves between reactants and products, with 'no catalyst' higher than 'with catalyst'
- upwards arrows to show E_a (from reactant level to tops of curves)

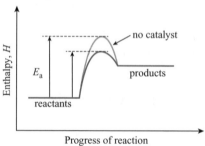

73. Dynamic equilibrium 1

The position of equilibrium moves to the left **(1)** in the direction of the greater number of molecules / moles of gas. **(1)**

74. Dynamic equilibrium 2

1 mark for each correct expression:

(a) $K_c = \dfrac{[HI(g)]^2}{[H_2(g)][I_2(g)]}$

(b) $K_c = \dfrac{[N_2O_4(g)]}{[NO_2(g)]^2}$

(c) $K_c = [H^+(aq)][OH^-(aq)]$

(d) $K_c = \dfrac{[H_2(g)][CO(g)]}{[H_2O(g)]}$

75. Industrial processes

(a) Low temperature **(1)** because the position of equilibrium will move to the right (in the direction of the exothermic reaction). **(1)**
High pressure **(1)** because the position of equilibrium will move in the direction of the fewest molecules of gas **(1)**.
A pressure of about 2 atm is used industrially as the rate of reaction is already very high at 1 atm.

(b) The temperature must be high enough to get a reasonable rate of reaction **(1)** without substantially reducing the equilibrium yield. **(1)**

77. Partial pressures and K_p

(a) $K_p = \dfrac{(p_{PCl3})(p_{Cl2})}{(p_{Cl5})}$ **(1)**

(b) $K_p = \dfrac{(p_{SO3})^2}{(p_{SO2})^2(p_{O2})}$ **(1)**

(c) $K_p = (p_{NH_3})(p_{HCl})$ **(1)**

78. Calculating K_c and K_p values

(a) $K_p = \dfrac{(p_{SO3})^2}{(p_{SO2})^2(p_{O2})}$ **(1)**

(b) $K_p = \dfrac{(1.7)^2}{(0.060)^2 \times (0.23)}$ **(1)**

$K_p = 3490$ **(1)** atm^{-1} **(1)**

79. Changing K_c and K_p

$K_p = \dfrac{(x_{NO_2})^2}{x_{N_2O_4}} \times \dfrac{(p_T)^2}{p_T} = \dfrac{(x_{NO_2})^2}{x_{N_2O_4}} \times p_T$ **(1)**

(to keep K_p constant when the total pressure is increased) the mole fraction / partial pressure of NO_2 must decrease / the mole fraction / partial pressure of N_2O_4 must increase. **(1)**

80. Acids, bases and pH

1 (a) H_2SO_4 because it is a proton donor / donates a proton to HNO_3. **(1)**

 (b) H_2SO_4 and HSO_4^-
    HNO_3 and $H_2NO_3^+$ **(1)**

2 pH $= -\log_{10}(2.5 \times 10^{-3})$
    $= -(-2.6) = 2.6$ **(1)**

81. pH of acids

(a) $[H^+] = 0.500 \, mol \, dm^{-3}$
    pH $= -\log_{10}(0.500) = 0.30$ **(1)**

(b) $[H^+] = \sqrt{K_a \times [acid]} = \sqrt{(1.35 \times 10^{-5}) \times 0.500}$
    $= 2.60 \times 10^{-3} \, mol \, dm^{-3}$ **(1)**
    pH $= -\log_{10}(2.60 \times 10^{-3}) = 2.59$ **(1)**

Both pH values must be to 2 decimal places, as this was asked for in the question.
Notice that the pH of the propanoic acid is almost the same as the pH of the ethanoic acid in the worked example, but propanoic acid (the weaker acid) is twice the concentration.

82. pH of bases

(a) $[H^+] = K_w/[OH^-]$
    $[H^+] = (1.00 \times 10^{-14})/0.500 \, mol \, dm^{-3}$
    $[H^+] = 2.00 \times 10^{-14} \, mol \, dm^{-3}$ **(1)**
    pH $= -\log_{10}(2.00 \times 10^{-14}) = 13.7$ **(1)**

(b) $[H^+] = K_w/[OH^-]$
    $[H^+] = (1.00 \times 10^{-14})/(2 \times 0.500) \, mol \, dm^{-3}$
    $[H^+] = 1.00 \times 10^{-14} \, mol \, dm^{-3}$ **(1)**
    pH $= -\log_{10}(1.00 \times 10^{-14}) = 14.0$ **(1)**
    Both pH values must be to 1 decimal place, as this was asked for in the question.

83. Buffer solutions

1 pH $= 4.76 + \log_{10}(0.400/0.200)$ **(1)**
    pH $= 5.06$ **(1)**
    The answer should be the same as the answer to the Worked Example.

2 [acid] $= 0.10 \div 0.25 = 0.40 \, mol \, dm^{-3}$
    [salt] $= 0.10 \div 0.25 = 0.40 \, mol \, dm^{-3}$
    Both for 1 mark
    pH $= 3.8 + \log_{10}(0.40/0.40) = 3.8$ **(1)**
    pH $= pK_a$ when equal concentrations of acid and salt are present. **(1)**

84. More pH calculations

$[H^+] = 10^{-9.65} = 2.24 \times 10^{-10} \, mol \, dm^{-3}$ **(1)**
$[NH_4^+]/[NH_3] = [H^+]/K_a = (2.24 \times 10^{-10})/(5.62 \times 10^{-10})$
           $= 0.4$ **(1)**
Alternatively, for first 2 marks
$pK_a = -\log_{10}(5.62 \times 10^{-10}) = 9.25$
$[NH_4^+]/[NH_3] = 10^{(9.25 - 9.65)} = 0.4$

Mix 0.4 volumes of ammonium chloride with 1 volume of ammonia solution. **(1)**

85. Titration curves

For each curve (shown here on two sets of axes for comparison) 1 mark for:

* appropriate starting pH
* appropriate pH at twice equivalence
* shape of curve, including near vertical section at the equivalence point with appropriate pH (weak acid > 7 and strong acid 7 into strong base; weak acid 7 and strong acid <7 into weak base)

Adding strong base:

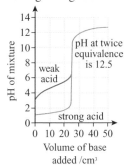

Adding weak base:

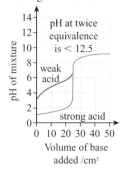

86. Determining K_a

1 $K_{In} = \dfrac{[H^+][In^-]}{[HIn]}$ **(1)**

 $[H^+] = K_{In} \times \dfrac{[HIn]}{[In^-]}$ **(1)**

 For the first colour change:
 $[H^+] = (1.2 \times 10^{-8}) \times 10 = 1.2 \times 10^{-7} \, mol \, dm^{-3}$ **(1)**
 pH $= 6.9$ **(1)**
 For the second colour change:
 $[H^+] = (1.2 \times 10^{-8}) \times 0.10 = 1.2 \times 10^{-9} \, mol \, dm^{-3}$ **(1)**
 pH $= 8.9$ **(1)**
 (Range is pH 6.9 – 8.9)

2 $K_a = 10^{-3.8}$
    $= 1.6 \times 10^{-4} \, mol \, dm^{-3}$ **(1)**

88. Born–Haber cycles 1

1 mark for each line with the correctly identified species, to 6 marks
1 mark for each correctly identified energy change, to 6 marks

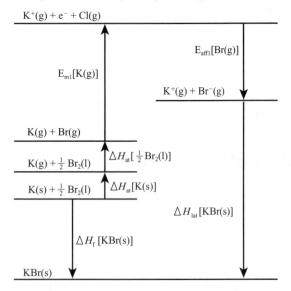

or

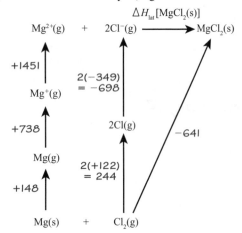

89. Born–Haber cycles 2

(a) Correct Born–Haber cycle, e.g.

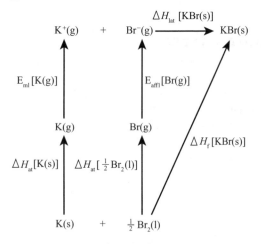

1 mark for each of the following:
- correct stages
- arrows in the correct directions
- all correct species
- all correct state symbols.

A Born–Haber cycle drawn as in part (b) of the Worked Example is also acceptable.
Enthalpy values are shown to help your understanding of part (b).

(b) $\Delta H_{latt} = -(+1451) - (+738) - (+148) - (-698) - (+244) + (-641)$ **(1)**
$= -1451 - 738 - 148 + 698 - 244 - 641$
$= -2524 \, kJ \, mol^{-1}$ **(1)**

90. An ionic model

Experimental lattice energy will be more negative. **(1)**
This means it is more exothermic.
Bromide ions are polarised (by the lithium ions) **(1)** leading to a degree of covalent bonding / degree of covalency / covalent character / deviation from pure ionic bonding. **(1)**

91. Dissolving

$\Delta_{sol}H[CaF_2] = -\Delta H_{lat}[CaF_2] + \Delta_{hyd}H[Ca^{2+}] + 2\Delta_{hyd}H[F^-]$ **(1)**
$= -(-2630) + (-1653) + 2(-474)$ **(1)**
$= +29 \, kJ \, mol^{-1}$ **(1)**

92. Entropy

1 The entropy will decrease / there will be a negative change. **(1)** (Bonds form between water molecules) so the molecules become more ordered / have fewer ways of being arranged **(1)** and their movement decreases. **(1)**

2 The entropy will increase/there will be a positive change because there are more moles of product than there are moles of reactant (and both are gases). **(1)**

93. Calculating entropy changes

(a) $\Delta H^{\ominus} = -241.8 - (-285.8) = +44.0 \, kJ \, mol^{-1}$ **(1)**
$\Delta S^{\ominus} = 188.7 - 69.9 = +118.8 \, J \, K^{-1} \, mol^{-1}$ **(1)**
$T = (44.0 \times 1000) \div 118.8$ **(1)**
$T = 370 \, K$ **(1)**
This is a minimum temperature for the change.

(b) $\Delta H^{\ominus} = -291.8 - (-285.8) = -6.0 \, kJ \, mol^{-1}$ **(1)**
$\Delta S^{\ominus} = 48.0 - 69.9 = -21.9 \, J \, K^{-1} \, mol^{-1}$ **(1)**
$T = (-6.0 \times 1000)/(-21.9)$ **(1)**
$T = 274 \, K$ **(1)**
This is a maximum temperature for the change.

94. Gibbs energy and equilibrium

(a) $-\Delta G/RT = \dfrac{-(13.9 \times 1000)}{(8.31 \times 500)} = -3.345$ **(1)**

$K = e^{-3.345} = 0.0353$ (no units) **(1)**

(b) The value for K is less than 1 so the position of equilibrium lies to the left **(1)** favouring reactants/N_2 and H_2. **(1)**

95. Redox and standard electrode potential

(a) Equation: $H^+(aq) + e^- \rightleftharpoons \frac{1}{2}H_2(g)$ **(1)**

1 mark for each of the following to 4 marks maximum:
- hydrogen gas / $H_2(g)$
- at 100 kPa
- (bubbled through) $1.0 \, mol \, dm^{-3}$
- hydrochloric acid / HCl(aq) / $H^+(aq)$
- at 298 K
- (over) a platinum electrode.

(b) It is defined as 0 V / it is 0 V by definition. **(1)**

96. Measuring standard emf

1 $Pt(s)|H_2(g)|H^+(aq)::Fe^{3+}(aq),Fe^{2+}(aq)|Pt(s)$ **(1)**

2 $E^{\ominus}_{cell} = E^{\ominus}_{right} - E^{\ominus}_{left}$
$= -0.76 - (-2.37) = +1.61 \, V$ **(1)**

97. Predicting reactions

(a) A reaction will occur:
$Zn(s) + CuSO_4(aq) \rightarrow ZnSO_4(aq) + Cu(s)$ **(1)**
$E^{\ominus}_{cell}$ is $+1.10\,V$/$E^{\ominus}$ for Cu^{2+}/Cu system is more positive. **(1)**

(b) A reaction will not occur: $E^{\ominus}_{cell}$ is $-0.34\,V$. **(1)**

98. Limitations of predictions

E_{cell} will decrease **(1)**
The position of equilibrium $Cu^{2+}(aq) + 2e^- \rightleftharpoons Cu(s)$ moves to the left. **(1)**
More electrons are released / right hand electrode becomes more negative (less positive). **(1)**

99. Storage cells and fuel cells

$Cd(OH)_2 + 2e^- + 2Ni(OH)_2 + 2OH^-$
$\rightarrow Cd + 2OH^- + 2NiO(OH) + 2H_2O + 2e^-$ **(1)**
but
$Cd(OH)_2 + 2Ni(OH)_2 \rightarrow Cd + 2NiO(OH) + 2H_2O$ **(2)**

101. Redox titrations

Amount of $MnO_4^- = 0.003\,00 \times 26.50 \times 10^{-3}$
$= 7.95 \times 10^{-5}\,mol$ **(1)**
Amount of Fe^{2+} in $25.0\,cm^3 = 5 \times 7.95 \times 10^{-5}$
$= 3.975 \times 10^{-4}\,mol$ **(1)**
Amount of Fe^{2+} in $250\,cm^3 = 10 \times 3.975 \times 10^{-4}$
$= 3.975 \times 10^{-3}\,mol$ **(1)**
Total mass of $Fe^{2+} = 3.975 \times 10^{-3} \times 55.8 = 0.2218\,g$ **(1)**
Percentage of iron $= 100 \times 0.2218 \div 0.25 = 88.7\%$ **(1)**

102. d-block atoms and ions

1 $1s^2\,2s^2\,2p^6\,3s^2\,3p^6\,3d^2$ **(1)**
2 $1s^2\,2s^2\,2p^6\,3s^2\,3p^6$ **(1)**
The Sc^{3+} ion has no electrons in d-orbitals. **(1)**
(To be a transition metal) it must form at least one ion with an incompletely filled d-orbital. **(1)**

103. Ligands and complex ions

(1)

The ethanedioate ion has two lone pairs of electrons available to form dative bonds with a metal ion. **(1)**

104. Shapes of complexes

(a) $[Cr(NH_3)_6]^{3+}$ **(1)**
hexaamminechromium(III) **(1)**
octahedral **(1)**

(b) $[MnCl_4]^{2-}$ **(1)**
tetrachloromanganate(II) **(1)**
tetrahedral **(1)**

105. Colours

The chloride (chloro) ligands split the d-orbital energies. **(1)**
d-d electron transitions absorb all light except for green light. **(1)**

106. Colour changes

They have different ligands **(1)** so the energy levels of the d-orbitals (in the central metal ion / iron(II) complex) are split by different amounts. **(1)**

107. Vanadium chemistry

VO_2^+ **(1)**
$V^{2+} + 2H_2O \rightarrow VO_2^+ + 4H^+ + 3e^-$ **(1)**
Manganate(VII) ions can oxidise vanadium(II) to vanadium(V):
$V^{2+} \rightarrow V^{3+} \rightarrow VO^{2+} \rightarrow VO_2^+$

108. Chromium chemistry

(a) $Cr_2O_7^{2-} + 14H^+ + 4Zn \rightarrow 2Cr^{2+} + 7H_2O + 4Zn^{2+}$
1 mark for correct species, 1 mark for correct balancing.

(b) $Cr^{2+}(aq)$ ions are blue and Cr^{3+} ions are green. **(1)**
Oxygen in the air oxidises Cr^{2+} ions to Cr^{3+} ions. **(1)**

109. Reactions with hydroxide ions

$[Al(H_2O)_6](aq) + 3OH^-(aq) \rightarrow [Al(H_2O)_3(OH)_3](s) + 3H_2O(l)$ **(1)**
$[Al(H_2O)_3(OH)_3](s) + OH^-(aq) \rightarrow [Al(H_2O)_2(OH)_4]^-(aq) + H_2O(l)$ **(1)**
in excess alkali

110. Reactions with ammonia

$[Al(H_2O)_3(OH)_3](s) + 3H_3O^+(aq) \rightarrow [Al(H_2O)_6]^{3+}(aq) + 3H_2O(l)$ **(1)**

111. Ligand exchange

There is an increase in the number of particles (from two to seven). **(1)**
The entropy change of the <u>system</u> is positive. **(1)**
There is a greater increase in stability when a monodentate ligand is replaced by a multidentate ligand than when it is replaced by a bidentate ligand.

112. Heterogeneous catalysis

Silver: adsorption is too weak / reactants not together long enough for a reaction. **(1)**
Tungsten: products are not desorbed from the surface. **(1)**

113. Homogeneous catalysis

1 They only have one oxidation number / do not have variable oxidation numbers. **(1)**
2 Fe^{2+} ions catalyse the reaction, but are not a product of the reaction. **(1)**

115. Measuring reaction rates

Method: measure the volume of hydrogen produced using a gas syringe / upturned measuring cylinder. **(1)**
Reasons *(two from the following for 1 mark each)*:
- Hydrogen is not very soluble (so will collect in the syringe / measuring cylinder).
- Hydrogen is not dense enough to measure the mass change easily.
- Magnesium chloride dissolves (to form a colourless solution).

116. Rate equation and initial rate

(a) If you compare experiments 1 and 4, [R] doubles while [P] and [Q] stay the same, and the rate stays the same. **(1)**
(b) $[P] = 0.25 \div 2 = 0.125\,mol\,dm^{-3}$
$[Q] = 0.60 \div 2 = 0.30\,mol\,dm^{-3}$
Both new concentrations for 1 mark ([R] can be ignored as it is zero order).
rate $= k[P][Q]^2 = 0.067 \times 0.125 \times 0.30^2$ **(1)**
$= 7.5 \times 10^{-4}$ **(1)** $mol\,dm^{-3}\,s^{-1}$

117. Rate equation and half-life

$k = \dfrac{0.693}{t_{\frac{1}{2}}} = \dfrac{0.693}{100} = 6.93 \times 10^{-3}$ **(1)** s^{-1} **(1)**

118. Rate-determining steps

Two molecules of NO_2 (as rate = $k[NO_2][NO_2]$) **(1)**
No molecules of CO (as CO is not in the rate equation) **(1)**

119. Finding activation energy

(a) (i) 0.00336 or 3.36×10^{-3} (K^{-1}) **(1)**
   (ii) −5.63 **(1)**
(b) E_a = $3250 \times 8.31/1000$ **(1)**
   = $+27.0 \, kJ \, mol^{-1}$ **(1)**

121. Identifying aldehydes and ketones

One isomer must be an aldehyde (propanal) and the other must be a ketone (propanone). **(1)**
1 mark for test used, 1 mark for correct observations, e.g.

Test	Result with aldehyde	Result with ketone
(warm with) acidified $K_2Cr_2O_7$	Orange solution turns green	No visible change / stays orange solution
Fehling's solution/ Benedict's solutions	Red precipitate forms	No visible change / stays blue solution
Tollens' reagent	Silver mirror forms	No visible change / stays colourless solution

122. Optical isomerism

1 mark for each diagram, which should clearly show object and mirror image. For example:

123. Optical isomerism and reaction mechanisms

C **(1)**

124. Reactions of aldehydes and ketones

Carbonyl compound: butanone **(1)** *It cannot be an aldehyde.*
Product with $LiAlH_4$: butan-2-ol **(1)** *Ketones are reduced to secondary alcohols.*
Product with HCN: 2-hydroxy-2-methylbutanenitrile **(1)** *This will be produced as a racemic mixture.*

125. Carboxylic acids

Oxidation of 3-methylbutan-1-ol using acidified (with dil H_2SO_4) $K_2Cr_2O_7$ **(1)**
$(CH_3)_2CHCH_2CH_2OH + 2[O] \rightarrow (CH_3)_2CHCH_2COOH + H_2O$ **(1)**
Oxidation of 3-methylbutanal using acidified (with dil H_2SO_4) $K_2Cr_2O_7$ **(1)**
$(CH_3)_2CHCH_2CHO + [O] \rightarrow (CH_3)_2CHCH_2COOH$ **(1)**
Hydrolysis of 3-methylbutanenitrile by refluxing with dilute hydrochloric acid **(1)**
$(CH_3)_2CHCH_2CN + H^+ + 2H_2O \rightarrow (CH_3)_2CHCH_2COOH + NH_4^+$
or $(CH_3)_2CHCH_2CN + HCl + 2H_2O \rightarrow (CH_3)_2CHCH_2COOH + NH_4Cl$ **(1)**
Structures could begin $CH_3CH(CH_3)CH_2$ instead of $(CH_3)_2CHCH_2$.

126. Making esters

1 propyl pentanoate **(1)** hydrogen chloride **(1)**
   1 mark for displayed formula:

2 *Two from the following for 1 mark each:*
   • Carbonyl C atom (in acyl chlorides) is more positive (than in carboxylic acids).
   • C–Cl bond is weaker than the C–O bond.
   • Cl^- ion is a better leaving group (than ^-OH).

127. Making polyesters

(a) Propane-1,3-diol **(1)**
   Butanedioic acid **(1)**
(b) condensation polymerisation **(1)**
   water **(1)**

128. Benzene

1 There are six carbon atoms in a cyclic compound **(1)**
   with three C=C bonds alternating with C–C bonds. **(1)**
2 six **(1)**

129. Halogenation of benzene

(a) electrophilic substitution **(1)**
   Both words needed for the mark.
(b) $AlCl_3 + Cl_2 \rightarrow [AlCl_4]^- + Cl^+$ **(1)**
   $[AlCl_4]^- + H^+ \rightarrow AlCl_3 + HCl$ **(1)**
(c) *1 mark for each curly arrow; one mark for correct intermediate, e.g.*

130. Nitration of benzene

(a) electrophilic substitution **(1)**
   Both words needed for the mark.
(b) $HNO_3 + H_2SO_4 \rightarrow NO_2^+ + HSO_4^- + H_2O$ **(1)**
(c) *1 mark for each curly arrow; 1 mark for correct intermediate, e.g.*

131. Friedel–Crafts reactions

(a) 1-chloropropane **(1)** $CH_3CH_2CH_2Cl$ **(1)**
   1 mark for displayed formula

   Can be shown in reverse.
(b) Propanoyl chloride **(1)** CH_3CH_2COCl **(1)**
   1 mark for displayed formula

   Can be shown in reverse.

132. Making amines

1 (a) Bromoalkane: 1-bromobutane (1)
   Nitrile: butanenitrile (1)
   (b) $CH_3CH_2CH_2CH_2Br + 2NH_3$
   $\rightarrow CH_3CH_2CH_2CH_2NH_2 + NH_4Br$ (1)
   $CH_3CH_2CH_2CN + 4[H] \rightarrow CH_3CH_2CH_2CH_2NH_2$ (1)
   (c) Dibutylamine is a secondary amine. (1)
   Butylamine acts as a nucleophile. (1)
   Further substitution happens (if ammonia is not in excess). (1)
2 Benzene has a delocalised π bond / rings of delocalised electrons. (1)
   These would repel the lone pair of electrons on ammonia / the nitrogen atom. (1)

133. Amines as bases

1 Answer is in the range 10.80 – 10.90. (1)
   Actual value is 10.84.
   The propyl group is more electron-releasing than the ethyl (and methyl) group. (1)
   The electron density on the nitrogen atom is greater / the lone pair of electrons become more available (to form a dative bond). (1)
2 The amino group / NH_2 group is not directly attached to the benzene ring. (1)
   The lone pair of electrons on the nitrogen atom cannot delocalise. (1)
   The methyl group is electron-releasing / makes the lone pair more available. (1)

134. Other reactions of amines

(a) $CH_3CH_2NH_2 + CH_3CH_2Cl$
   $\rightarrow CH_3CH_2NHCH_2CH_3 + HCl$ (1)
(b) $CH_3CH_2NHCH_2CH_3 + CH_3CH_2Cl$
   $\rightarrow CH_3CH_2N(CH_2CH_3)_2 + HCl$ (1)
(c) $CH_3CH_2N(CH_2CH_3)_2 + CH_3CH_2Cl$
   $\rightarrow CH_3CH_2N(CH_2CH_3)_3^+ + Cl^-$ (1)

135. Making amides

(a) $CH_3CH_2COCl + NH_3 \rightarrow CH_3CH_2CONH_2 + HCl$ (1)
   propanamide (1)
(b) $CH_3CH_2COCl + CH_3CH_2CH_2NH_2$
   $\rightarrow CH_3CH_2CONHCH_2CH_3 + HCl$ (1)
   N-propylpropanamide (1)

136. Making polyamides

(a) Octane-1,8-dioic acid:

(1)

Hexane-1,6-diamine:

(1)

(b) Repeat unit:

(1)

(c) The 6 refers to the number of carbon atoms in the diamine and the 8 refers to the number of carbon atoms in the dicarboxylic acid. (1)

137. Amino acids

1 In alkaline solution:

(1)

In acidic solution:

(1)

2 2-amino-3-hydroxypropanoic acid (1)

138. Proteins

glycine–cysteine:

(1)

cysteine–glycine:

(1)

140. Grignard reagents

Butan-1-ol (1)
$CH_3CH_2CH_2CH_2OH$ (1)

141. Methods in organic chemistry

1 The product is more soluble in hot solvent / the solvent is saturated with the product. (1)
   (The product is less soluble in cold solvent so) excess product forms as mixture is cooled. (1)
2 *Two suitable hazards with appropriate control measures, 1 mark for each hazard and 1 mark for its control measure (which must match the hazard), e.g.*
   • Ethanol is flammable / avoid contact with naked flames.
   • Ethanol is harmful / wear eye protection.
   • Ethanal produces harmful fumes / ensure adequate ventilation.
   • Potassium dichromate(VI) is toxic or oxidising / wear gloves and eye protection.
   • Dilute sulfuric acid is irritant or corrosive / wear gloves and eye protection.
   • Naked flames may cause fires / tie hair back.

142. Methods in organic chemistry 2

(a) Reflux (1)
(b) Steam distillation (1)

143. Reaction pathways

*1 mark for each labelled pathway to a maximum of **6** marks, e.g.*

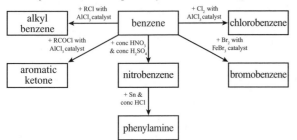

There are many other ways this chart could be presented.

144. Chromatography

$R_f = 0.82$ **(1)**
Value accepted 0.80–0.85, and show working out.

145. Functional group analysis

Answers should include the following points in a logical sequence:
- Warm each alcohol with potassium dichromate(VI) **(1)**
- acidified with dilute sulfuric acid **(1)**
- under distillation conditions. **(1)**
- The mixture stays orange with the tertiary alcohol/only turns green with the primary and secondary alcohols. **(1)**
- Test the <u>products</u> with Fehling's / Benedict's / Tollens'. **(1)**
- The product from the secondary alcohol should give no change / only the product from the primary alcohol will produce a change. **(1)**

This works because primary alcohols are oxidised to aldehydes and secondary alcohols are oxidised to ketones (if reflux conditions are used, primary alcohols are oxidised further to carboxylic acids).

146. Combustion analysis

Mass of C = $1.76 \times 12.0 \div 44.0 = 0.48\,g$ **(1)**
Mass of H = $0.72 \times 2 \times 1.0 \div 18.0 = 0.08\,g$ **(1)**
Mass of O = $0.88 - (0.48 + 0.08) = 0.32\,g$ **(1)**

C	H	O	
$\dfrac{0.48}{12.0} = 0.40$	$\dfrac{0.08}{1.0} = 0.080$	$\dfrac{0.32}{16.0} = 0.020$	**(1)**
$0.040 \div 0.020 = 2$	$0.080 \div 0.020 = 4$	$0.020 \div 0.020 = 1$	**(1)**

Empirical formula is C_2H_4O. **(1)**

147. High-resolution mass spectra

M_r of CH_3COCH_3
$\quad = (3 \times 12.0107) + (6 \times 1.0078) + (1 \times 15.9994)$
$\quad = 36.0321 + 6.0468 + 15.9994 = 58.0783$ **(1)**
M_r of $H_2NCH=CHNH_2$
$\quad = (2 \times 12.0107) + (6 \times 1.0078) + (2 \times 14.0067)$
$\quad = 24.0214 + 6.0468 + 28.0134 = 58.0816$ **(1)**
Y is likely to be propanone **(1)** because its M_r is closest to the M_r for propanone. **(1)**

148. ^{13}C NMR spectroscopy

(a) One peak **(1)** because all six carbon atoms are in the same chemical environment / are equivalent. **(1)**
(b) Four peaks **(1)** explanation, e.g.

$$\begin{array}{c} 1\ CH_3 \\ | \\ H_3C - CH - CH_2 - CH_3 \\ \ \ \ \ 1\ \ \ \ 2\ \ \ \ \ \ 3\ \ \ \ \ \ 4 \end{array}$$ **(1)**

149. Proton NMR spectroscopy

1 (a) 1 **(1)**
   (b) 1 **(1)**
   (c) 2 **(1)**
2 (a) 2 peaks
      The hydrogen atoms in the three CH_3 groups are in the same chemical environment / equivalent **(1)** (and produce one peak).
      The hydrogen atoms in the right hand CH_3 group produce 1 peak. **(1)**
   (b) 9 hydrogen atoms in the three left hand CH_3 groups, 3 in the right hand CH_3 group. **(1)**
   (c) 3 : 1 **(1)**
      This shows that the peak area ratio tells you the relative numbers of hydrogen atoms in each environment and not always their actual numbers.

150. Splitting patterns

CH_3 group: triplet as it is next to a CH_2 group **(1)**
Left hand CH_2 group: quartet as it is next to a CH_3 group **(1)**
Right hand CH_2 group: doublet as it is next to a CH group **(1)**
CH group: triplet as it is next to a CH_2 group **(1)**
OH group: singlet as it is not adjacent to non-equivalent hydrogen atoms. **(1)**

Notes

Published by Pearson Education Limited, 80 Strand, London, WC2R 0RL.

www.pearsonschoolsandfecolleges.co.uk

Copies of official specifications for all Edexcel qualifications may be found on the website: www.edexcel.com

Text and illustrations © Pearson Education Limited 2015
Copyedited by Hilary Herrick and Marilyn Grant
Typeset and illustrations by Tech-Set Ltd, Gateshead
Produced by Out of House Publishing
Cover illustration by Miriam Sturdee

The right of Nigel Saunders to be identified as author of this work has been asserted by him in accordance with the Copyright, Designs and Patents Act 1988.

First published 2015

18 17
10 9 8 7 6 5 4 3

British Library Cataloguing in Publication Data
A catalogue record for this book is available from the British Library

ISBN 978 1 447 98997 4

Printed in Slovakia by Neografia

Acknowledgements
The publisher would like to thank the following for their kind permission to reproduce their photographs:
(Key: b-bottom; c-centre; l-left; r-right; t-top)
Pearson Education Ltd: 26, Beehive Illustration / Mark Turner; 47, Science Photo Library Ltd; Martyn F. Chillmaid
All other images © Pearson Education

A note from the publisher
In order to ensure that this resource offers high-quality support for the associated Pearson qualification, it has been through a review process by the awarding body. This process confirms that this resource fully covers the teaching and learning content of the specification or part of a specification at which it is aimed. It also confirms that it demonstrates an appropriate balance between the development of subject skills, knowledge and understanding, in addition to preparation for assessment.

Endorsement does not cover any guidance on assessment activities or processes (e.g. practice questions or advice on how to answer assessment questions), included in the resource nor does it prescribe any particular approach to the teaching or delivery of a related course.

While the publishers have made every attempt to ensure that advice on the qualification and its assessment is accurate, the official specification and associated assessment guidance materials are the only authoritative source of information and should always be referred to for definitive guidance.

Pearson examiners have not contributed to any sections in this resource relevant to examination papers for which they have responsibility.

Examiners will not use endorsed resources as a source of material for any assessment set by Pearson.

Endorsement of a resource does not mean that the resource is required to achieve this Pearson qualification, nor does it mean that it is the only suitable material available to support the qualification, and any resource lists produced by the awarding body shall include this and other appropriate resources.